IMAGES
of America

INYO NATIONAL FOREST

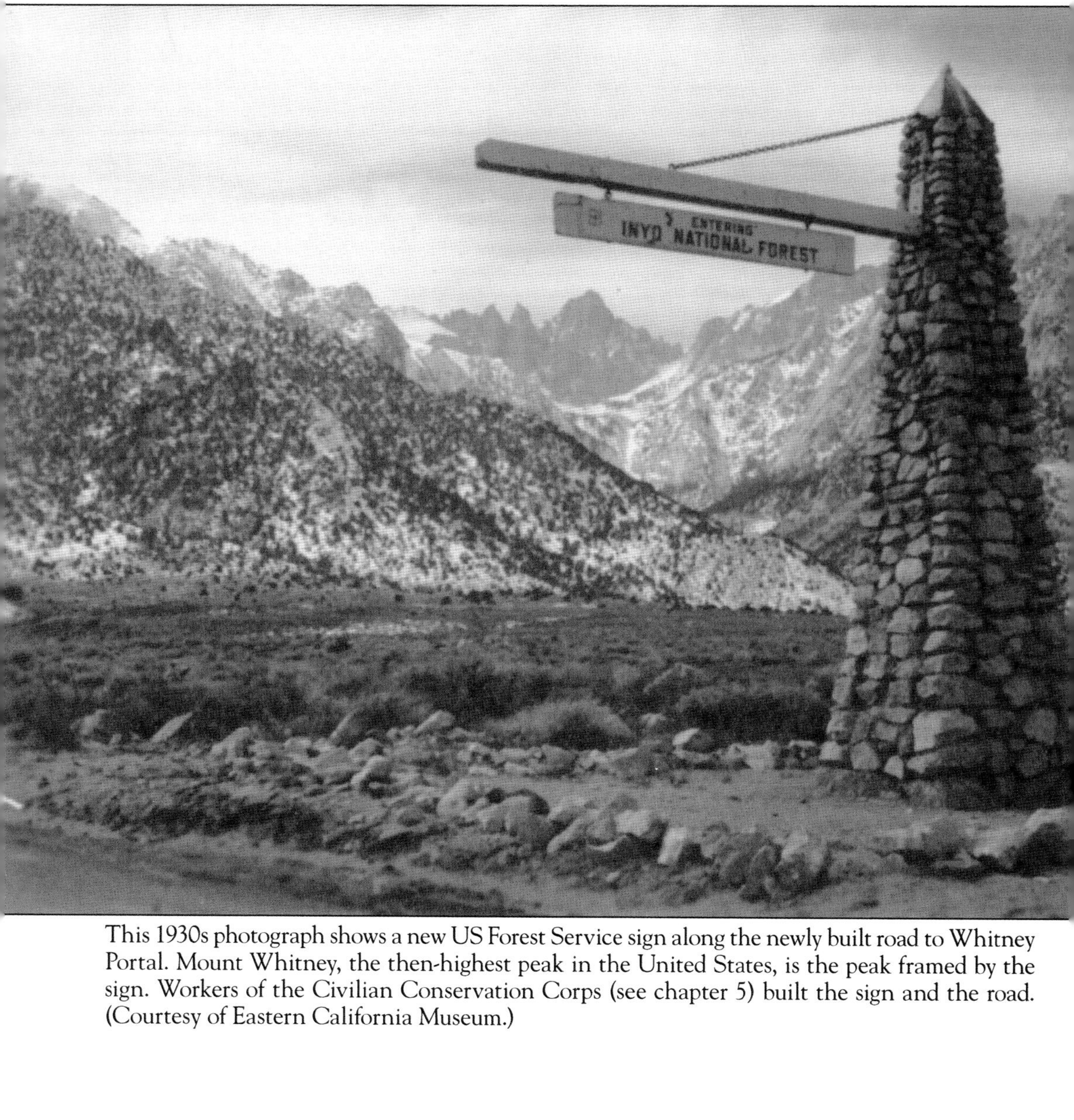

This 1930s photograph shows a new US Forest Service sign along the newly built road to Whitney Portal. Mount Whitney, the then-highest peak in the United States, is the peak framed by the sign. Workers of the Civilian Conservation Corps (see chapter 5) built the sign and the road. (Courtesy of Eastern California Museum.)

On the Cover: A mid-1920s camping scene is shown here near the shore of Saddlebag Lake, located at the northern end of the Inyo National Forest. At 10,050 feet above sea level, 3.5 miles north of Tioga Pass, Saddlebag Lake was originally a smaller, natural lake, named "Saddlebags" for its twinned shape. In 1922, the California-Nevada Power Company expanded the lake with a storage dam for a hydroelectric plant in Lee Vining Canyon, 2,250 feet below. Today, Saddlebag Lake is still popular for camping, fishing, and hiking. (Photograph by Burton Frasher, courtesy of Eastern California Museum.)

IMAGES
of America

INYO NATIONAL FOREST

Andy Selters for the
Eastern Sierra Interpretive Association

ARCADIA
PUBLISHING

ISBN 978-0-7385-9305-0

Published by Arcadia Publishing
Charleston, South Carolina

Printed in the United States of America

Library of Congress Control Number: 2011936839

For all general information, please contact Arcadia Publishing:
Telephone 843-853-2070
Fax 843-853-0044
E-mail sales@arcadiapublishing.com
For customer service and orders:
Toll-Free 1-888-313-2665

Visit us on the Internet at www.arcadiapublishing.com

For Ace, who I hope will have as full a life
with the Sierra Nevada as I have

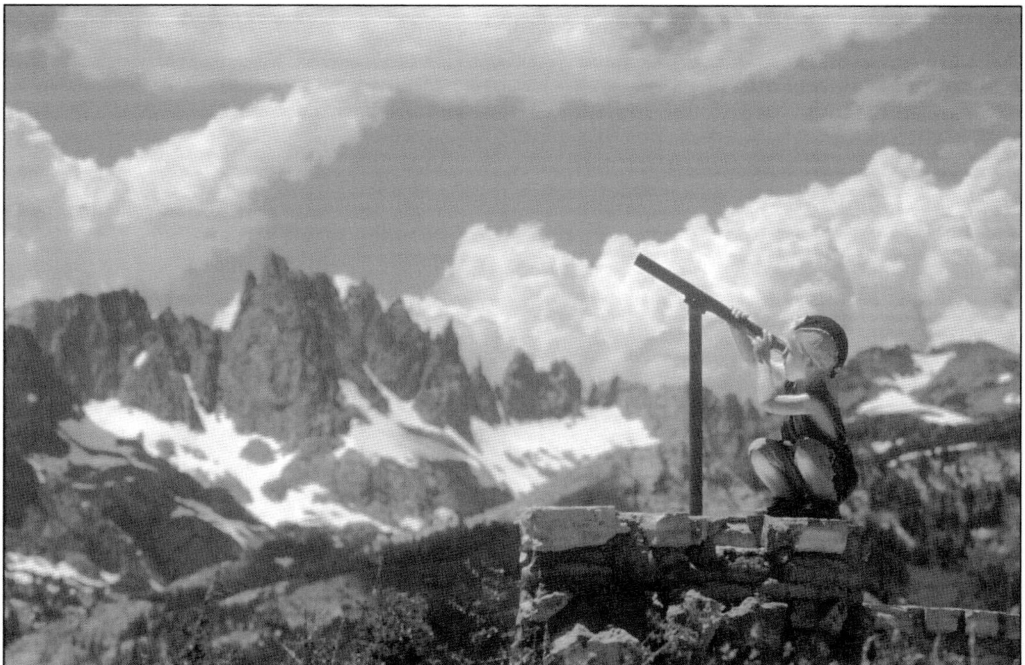

Ace Selters, age four, checks out the view at Minaret Summit. (Courtesy of AS.)

Contents

Acknowledgments

Many people whose lives are embedded in the Eastern Sierra contributed to this book. Lisa Isaacs pointed me to the project, and it was Debbe Eilts, director of the Eastern Sierra Interpretive Association (ESIA), whose inspiration and oversight made this possible. Nancy Upham, public affairs officer on the Inyo, found time in her overwhelming schedule to open the Forest's doors for me and applied her long experience to vital fact checking of the text. Heritage resources specialist Sarah Johnston and her assistant Paul Hoornbeck graciously let me into the Forest's archives and their office space, and Diana Pietrasanta contributed historical notes, too. John Louth, manager of the Ancient Bristlecone Pine Forest, also came up with key photographs and facts. At the Eastern California Museum, director Jon Klusmire and archivist Roberta Harlan deserve more thanks than I can give; they proved that they have made the museum our region's premier historical resource. Paiute historian Gerald Howard gave me invaluable answers to my questions relating to his tribe. So many others who are vital parts of the history of the Inyo were generous with their stories and photographs, notably Lou and Marye Roeser, Phil Pister, Andrea Tetrick, John Boothe and Jill Kinmont-Boothe, Helen Dixon, and Rocky and Tish Rockwell. Robin Morning and her tremendous book *Tracks of Passion* had so much ski history to share that it is a shame I couldn't use more. The following people all gave of their time, tales, and photographic archives: Wynne Benti of Coon's Gallery, Pam Vaughn of the Laws Railroad Museum, Chris Langley of the Lone Pine Film History Museum, Jeff McCarthy of Southern California Edison (SCE), Barbara Laughon and Kevin Calder of Mountain Light Gallery, Laurie Williams, Chuck and Marin Spencer, Ray Grey, Tom Ross, Chris Cox, Gordon Wiltsie, Gene Rose and Phil Bartholomew, Daniel Pritchett of White Mountain Research Station, Deanna Dulen of Devils Postpile National Monument, Jennifer and Lee Roeser of McGee Creek Pack Station, Ingrid Wicken of the California Ski Library, Frank Cassidy, Jeff and Maranda Moran, John Dittli, Jim Stimson, Jo Bacon, Mary Walker and Dan Hansen of Mammoth Mountain Ski Area, Arya Degenhardt of the Mono Lake Committee, Randy Gillespie, and Norm DeChambeau of the Mono Basin Historical Society. Overall, I sit in awe with how much I've learned through all these people and institutions and how much I couldn't fit into these covers.

Abbreviated photograph credits refer to the following: INF, Inyo National Forest; ECM, Eastern California Museum; MMSA, Mammoth Mountain Ski Area; SCE, Southern California Edison; WMRS, White Mountain Research Station; and AS, Andy Selters.

Foreword

Although most well-known today for its outstanding recreational opportunities, exquisite scenery, and inspiring destinations, for over 100 years, the lands of the Inyo National Forest have also provided rangeland for grazing cattle and sheep, mining opportunities, and even timber production. The watersheds have provided water for local communities as well as the city of Los Angeles.

Each year, the Inyo National Forest plays host to millions of visitors who come here to enjoy the spectacular scenery and the multitude of varied recreational opportunities that are found on this two-million-acre national forest.

People first started coming to actually recreate on the land known today as the Inyo National Forest as early as the 1860s. Although the types of recreation people engage in today are considerably different than the late 1800s and the first half of the last century, some things have remained constant. In 1937, Forest Supervisor Roy Boothe sought out the reasons why people visited the Inyo, and out of the over 500 people that were interviewed, the top draws were fishing, looking at the spectacular scenery, and rest and relaxation. We find much of the same today, with hiking, skiing, and exploring being way up on the charts as well.

In my time on the Inyo, I have found that the Eastern Sierra is a special place where people care deeply for their public lands and all that they have to offer. Although there is not always agreement on how these lands should be managed and used by those who play on them, are financially dependent on them, or live near them, all place a high value on the health and sustainability of the land.

With the many challenges that lie ahead of us, I am heartened to see the number of people who are stepping up to help as volunteers and community stewards of the Inyo National Forest. Today, we have strong partnerships with many organizations that are helping to raise public awareness about the need for responsibly recreating on our public lands as well as helping people to understand the value of protecting the sustainability of our resources and the need for ecosystem restoration. They are working hard to organize and supervise a multitude of resource and recreation volunteer projects where citizens from near and far are learning the joy of spending time helping to take care of the land.

I invite you to stop by one our visitor centers from Lone Pine to Mono Lake to learn more about the Inyo National Forest and its wonderful and rich history and to find out how you can become involved in helping to care for this awesome place known as the Inyo National Forest.

—Ed Armenta
Forest Supervisor
September 2011

INTRODUCTION

Along the eastern edge of California, the Sierra Nevada and the White and Inyo Mountains rise into some of America's most stunning and beloved terrain. For over 100 years, these mountains have been under federal management. They make up most of the Inyo National Forest. In that century, people have worked, studied, explored, and, especially, played here. For decades now, the Inyo has been one of the most-visited national forests in America, with visitor-ship comparable to the most popular national parks.

The primary feature of the Inyo is the eastern scarp of the Sierra. This is the highest and most extensive mountain front in the lower 48, a 275-mile-long fault block where summits tower up to 10,000 feet above the valley below, with eight peaks reaching over 14,000 feet above sea level. Paralleling the Sierra a dozen miles away, the Whites and Inyos are a 130-mile-long spine that also rises to over 14,000 feet.

The Inyo is a national forest of superlative grandeur but relatively little forest. The cover of pines and firs is generally sparse, and there are large areas with no trees at all. At the turn of the 20th century, these ranges were brought into federal protection mostly to ensure the delivery of clean water. In winter, the stormy heights collect a blanket of snow measured in the dozens of feet. Come summer, that bounty melts to fill sublime alpine lake basins and then tumbles down rugged canyons into the arid valleys below. When the great naturalist John Muir scanned these horizons from a Sierra summit, he wrote, "Eastward, the whole region seems a land of desolation covered with beautiful light."

Today, when newcomers arrive here along Highway 395, they typically do a double take: "Wow. This is a *landscape*." That experience has been a perpetual tale. From the native Americans through the miners, shepherds, rangers, loggers, fishers, hunters, campers, scientists, hydrologists, artists, road workers, skiers, movie-makers, and mountaineers who have come to know this land, thousands of people have told how this terrain has written itself into their hearts. These mountains have helped give America the concept of transcendence by geography and they are an iconic and vital part of the nation.

For at least 600 years, this area has been the home of 1,000 to 2,000 Paiutes. The term *Pai-ute* refers to Ute people who live along watercourses, but here they refer to themselves as *Numu*. They had a successful, seminomadic way of life, hunting deer, bighorn sheep, and small game, and gathering plant life, especially piñon nuts. Their basketry is still praised as some of the finest on the continent. They also crafted ceramic pots and did rudimentary cultivation. The Numu knew the Sierra well. They hunted or visited even to the summits and journeyed over the passes to trade salt and obsidian for acorns and other goods. While legend holds that the word *Inyo* is a Paiute word meaning "dwelling place of the great spirit," modern Paiutes do not know this word. One educated guess is that the name Inyo originated as a misinterpretation of the Spanish word *Indio*—translates to "Indian"—spoken by Mexicans referring to the mountains east of the Owens Valley, subsequently romanticized.

White men first showed up here in the 1850s. The Paiutes hid in the mountains to watch and hope that the strange, new people would leave. But of course, they did not, and the story of immigrant conquest oft repeated across the West played out. The Paiutes and settlers parried back and forth, the Paiutes attacking livestock with guerilla tactics and the settlers chasing back. With the help of rifles obtained from Americans farther north, the Paiutes for a while held on. Both sides dug in. But the US Army arrived and built a fort at Camp Independence. They rounded up and slaughtered many Paiutes, and in 1863, the rest were marched to a reservation at Fort Tejon. In time, their descendants returned to the area, and in 1940, the Bishop Reservation was established. Paiutes still regard the mountains as sacred terrain that holds the legacy of their people.

Even as those battles raged, prospectors started spilling over from the west side of the Sierra, and boomtowns and mining districts began to sprout. The first was "Dogtown," north of Mono Lake, and soon after came "Alabama" and "Kearsarge," farther south. Across the West, frontier frenzy took more than what could be sustained and did a lot of damage. By 1890, the US Census director fairly certified that the frontier was settled, and the nation took up the call to restrain what was essentially a free-for-all. Conservationists and Preservationists in the Progressive Movement fired up optimism for a sustainable Western abundance by preserving the most scenic terrain in national parks and by bringing other lands into public administration for science-based harvests of wood, forage, and water. Today, many ask why the spectacular Eastern Sierra was not made into a national park, but when the area's future was being decided, that option was never considered. Arid scenery was as yet unappreciated, and few Americans had scaled the intimidating east-side canyons to see the verdant high country.

In 1893, wholesale federal protection came to the Sierra for a while, when President Harrison "withdrew" the Sierra Forest Reserve. This was the second and ultimately the largest reserve designated under the 1891 Forest Reserve Act, which operated under the Department of Interior. The Sierra Reserve was formed especially out of the call from leaders in the San Joaquin Valley, who were living in the downstream legacy of rapacious logging, grazing, and hydraulic mining. The east side remained a hinterland with little or no federal presence.

In 1905, the administration of Theodore Roosevelt reassigned the Forest Reserves to a new agency within the Department of Agriculture, the US Forest Service. The move was a signal that Roosevelt wanted these lands for managed resource harvest. To steer a course between private interests (who wanted unbounded access) and preservationists including Muir, Roosevelt picked his right-hand resource man, Gifford Pinchot, to design and set up the new agency. Pinchot's life calling was to foster the efficient, sustainable harvesting of natural resources. He envisioned the Forest Service to become a hierarchical organization of academically trained professionals. He wrote, "All the resources of the forest reserves are for use, and this use must be brought about in a thoroughly prompt and business-like manner, under such restrictions as will ensure the permanence of these resources." Pinchot became the Forest Service's originating chief, and the Sierra Forest Reserve became the Sierra National Forest.

During these same years, leaders and barons in Los Angeles were plotting to turn their small city into a booming metropolis, using water from the faraway Owens River. Agents for the city had begun surreptitiously acquiring as much private land and water rights as they could along the Owens, but the city wanted assurance that federal lands would remain clear of homesteads that could take water or block future reservoir plans. As part of his agenda to counteract the period's corporate barons, Roosevelt himself took up the cause to see a coastal city blossom with a strong municipal water agency. On May 25, 1907, he signed an order creating the first incarnation of the Inyo National Forest. This was a strip of 221,324 acres along the Owens River above Bishop that was "withdrawn" exclusively to help ensure the river would be diverted to Los Angeles.

Protecting watersheds was an important national forest mandate, but exporting water far away made Owens Valley residents and many others argue that this overstepped the letter and spirit of the law. Controversy ran high and a legal challenge would have been interesting, but, as Roosevelt himself noted, the Owens Valley citizenry were simply too few to stand in the way of one of the largest public works projects in the nation's history. Five years later, Roosevelt's handpicked successor William Howard Taft removed most of the original Inyo lands from US Forest Service domain, but before any homestead claims were processed, the aqueduct was a fait accompli. In 1913, the aqueduct gates were opened, and the Owens River began a historic 223-mile-long journey to Southern California. In less than 20 years, the city's population would quintuple, and its county became the leading agricultural county in the nation. With little water available, the Owens Valley was nearly depopulated.

Meanwhile, the scale of the Sierra was making it clear that the vast Sierra National Forest needed to be divided into more manageable units. In 1908, Roosevelt faced congressional opposition to any expansion of public lands management and the whole national forest concept. In a "midnight" executive order, he created several dozen new national forests all over the West. His act also partitioned the Sierra's eastern scarp off into two national forests. A much grander Inyo National Forest now included the southern half of the Sierra scarp and most of the White-Inyo range, too. The northern half of the Eastern Sierra became the Mono National Forest.

The Forest Service continued to consolidate terrain into the Inyo's territory. In 1920, the Mount Whitney and Kern Plateau areas were transferred out of the Kern National Forest to the Inyo. After World War II, the Mono National Forest was merged into the Inyo. That filled out the general scope of the Inyo National Forest, as we now know it; although, occasional land exchanges continue to adjust the Inyo's boundaries to this day.

The Inyo's terrain and location and its position within the National Forest System are the defining forces of its history. Considering congressional budgets and federal oversight have focused notably on national forests with big timber, Inyo supervisors and staff historically have been left to manage more to their own devices and judgment. Meanwhile, the public has steadily found the Inyo to be a recreation wonderland, a fairly accessible place of national park character to fish, hunt, ride, camp, climb, ski, and more. Gradually, the Forest Service's national mandate expanded to prioritize the enhancement of recreation and the protection of wilderness resources, and the Inyo became one of the Forest Service's flagship national forests for non-timber goals. Throughout, the Inyo has continued to serve its original purpose as a protected watershed for Los Angeles, and it has collected many tales and devoted characters.

One

From Frontier to National Forest

Early Management

The earliest rangers on the Inyo were few and they must have roamed alone over long distances trying to encourage sheepmen and lumberjacks to be reasonable in their take. Within a few years, the federal presence increased, ranger stations were built, and more rangers and a forest supervisor were assigned.

In 1925, a young World War I veteran named Lawson Brainerd was hired as a district ranger and he left colorful and romantic portrayals of his job. When he first stepped off the train near Big Pine, he immediately fell in love with the terrain. "It would be a soul of iron who wouldn't thrill to the panorama . . . It was my country and always would be," said Brainerd.

At work, Brainerd described a typically chaotic morning as follows: "The Coyote Range Association were due to drive their cattle onto the Coyote Range [a high bench southwest of Bishop] and the critters had to be counted. Paul Lucco's sheep were nibbling off, and swallowing with gusto, tidbits of Parker's Oak Creek cattle range and Parker was squawking over the phone like a cheated harlot. Across the Owens Valley I could see the dust from two bands of nomadic sheep illegally slicing off generous hunks from the Inyo Mountain Range. Numerous letters from the R.O. [Regional Office, in San Francisco] demanding immediately the delinquent reports on Special Use Permits. Various requests for woodcutting from Independence to Big Pine. Local delegations offering unrequested advice on how the Forest Service should be run. . . . In my ignorance, all was as tangled as a barrel of well-shaken clothes hangers."

In addition, the exploding population of Los Angeles already was beginning to make the Inyo their favorite getaway destination. Though it was a two-day drive from Los Angeles to Bishop, as early as August 1916, ranger Louis Barrett remarked that recreation "travel was just starting to get heavy."

Early Inyo staff had to be outdoorsmen and diplomats well endowed with make-do common sense. In 1926, a man with only an eighth-grade education but can-do skills in spades became the supervisor. For 19 years, Roy Boothe would shepherd the agency into a respected and effective organization.

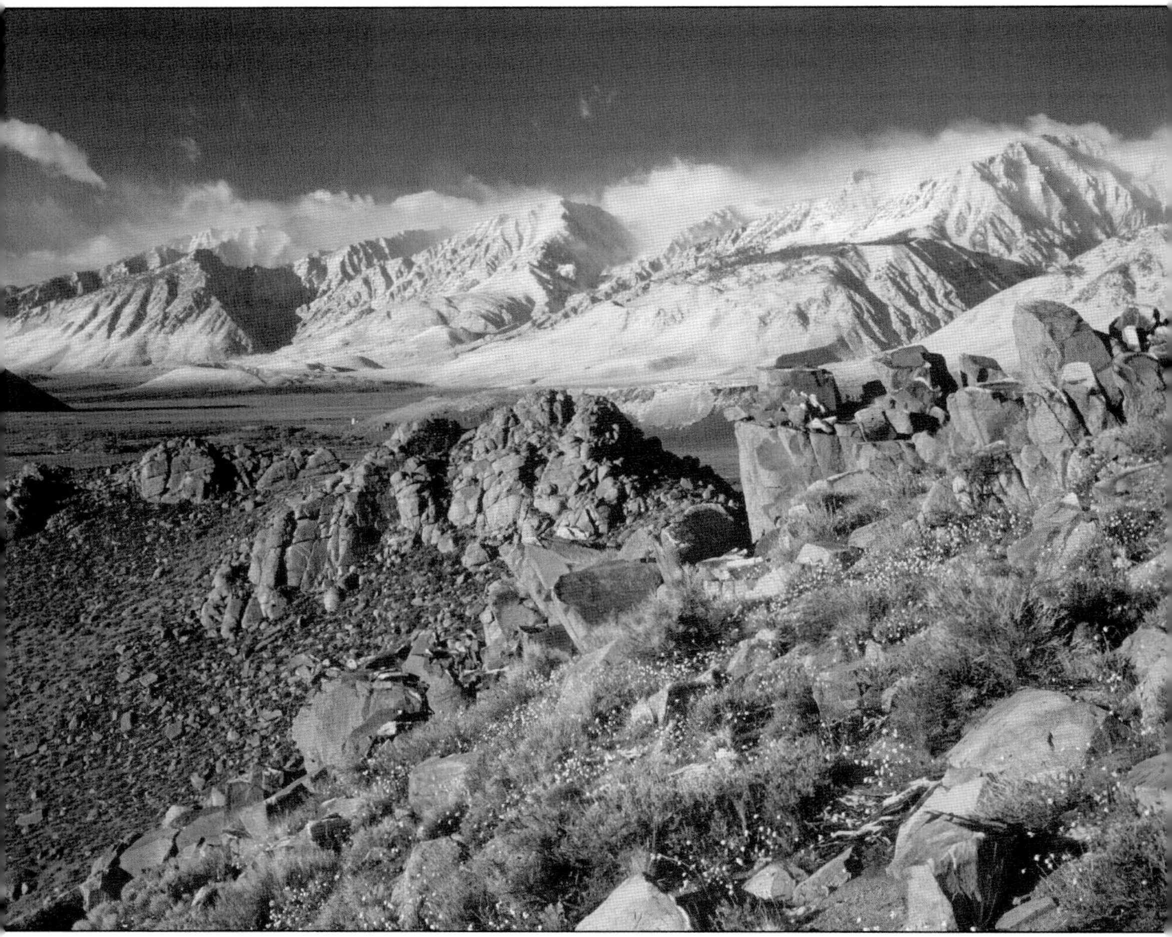

Here is the Sierra's eastern scarp in the clearing of a spring 2007 snowstorm, looking south from north of Sawmill Canyon. The ridges along the far left rise to Mount Williamson, which is 14,375 feet high. The boundary of the Inyo National Forest generally runs at about 5,000 feet along and a bit out from the foot of the mountains and up to the watershed crest. Through the 19th century, this scarp was seen mostly as the harsh, intimidating side of the Sierra, a barrier to travel. As sheepmen and miners began to make use of the mountains and then as trails were built up many of the canyons, perceptions gradually softened. Now, this more dramatic side is for many people their favorite side of the Sierra. Federal and regional political processes at the dawn of the 20th century assigned about 95 percent of both Inyo and Mono Counties here as public land. (Courtesy of AS.)

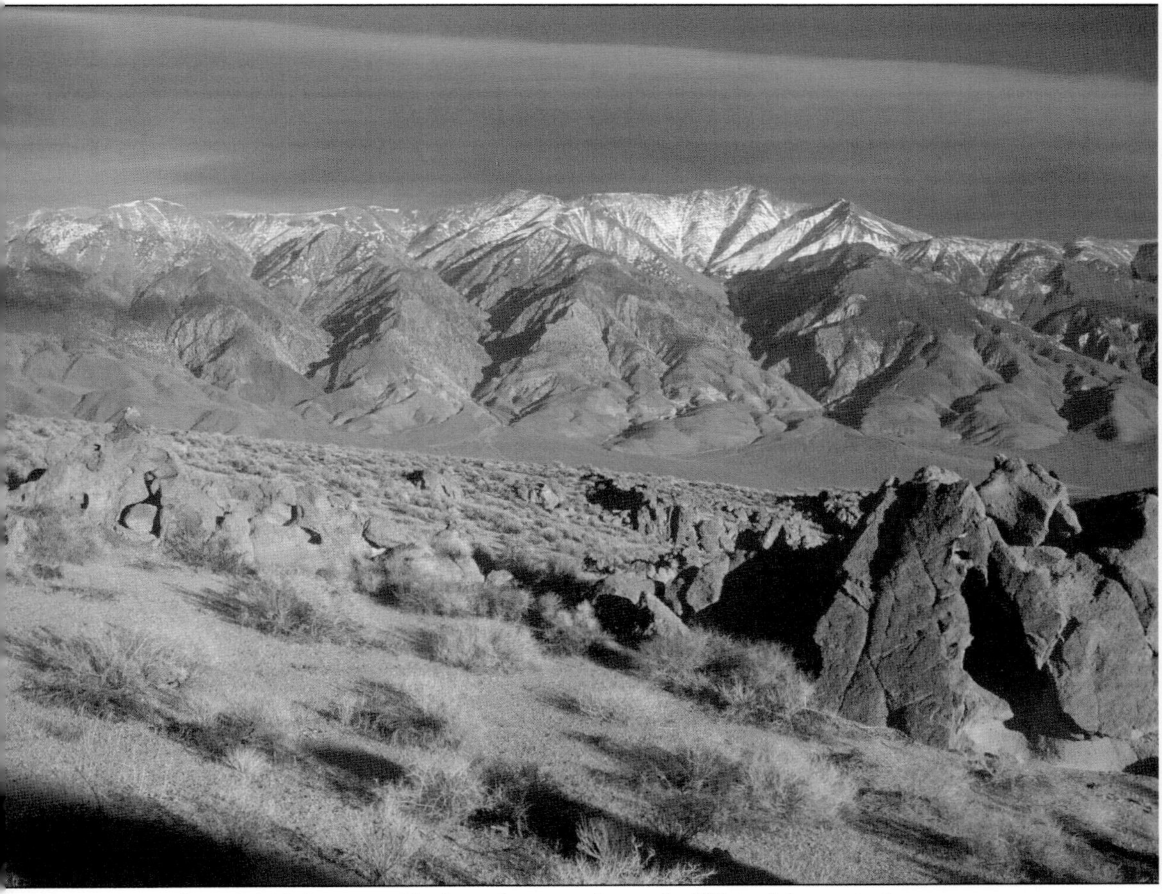

As seen from about 5,000 feet near Chidago Canyon, this is the White Mountains' western scarp. The high point is White Mountain Peak at 14,246 feet high. The range was named for the white granite at its northern end. Mount Barcroft is just out of view to the right. The rolling crest of the Whites extends northwest (left) at over 13,000 feet past Pelliser Flats, the most extensive span of continuous alpine terrain in the lower 48. To the south the range continues, with a break at 7,000-foot-high Westgard Pass, into the Inyo Mountains. (Courtesy of AS.)

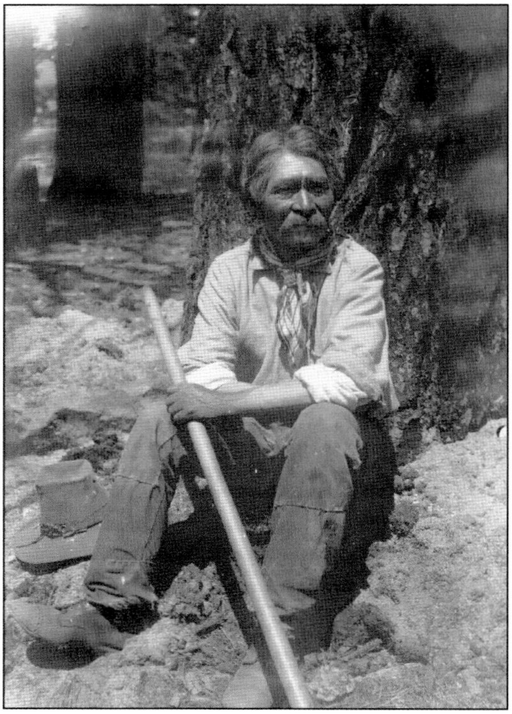

Sim Lundy was a Mono Basin Paiute who, like many Paiutes, took a white name to function in the new world and probably kept a native name to himself. Lily LaBraque, author of *Man from Mono*, wrote that Lundy became her friend and that he suffered a crippling leg injury from a horse as a teenager. (Courtesy of INF.)

Here is a Paiute woman at her Owens Valley home, called a *toni,* in 1903. Note the tumpline and frame for carrying her load. (A.A. Forbes photograph, courtesy of ECM.)

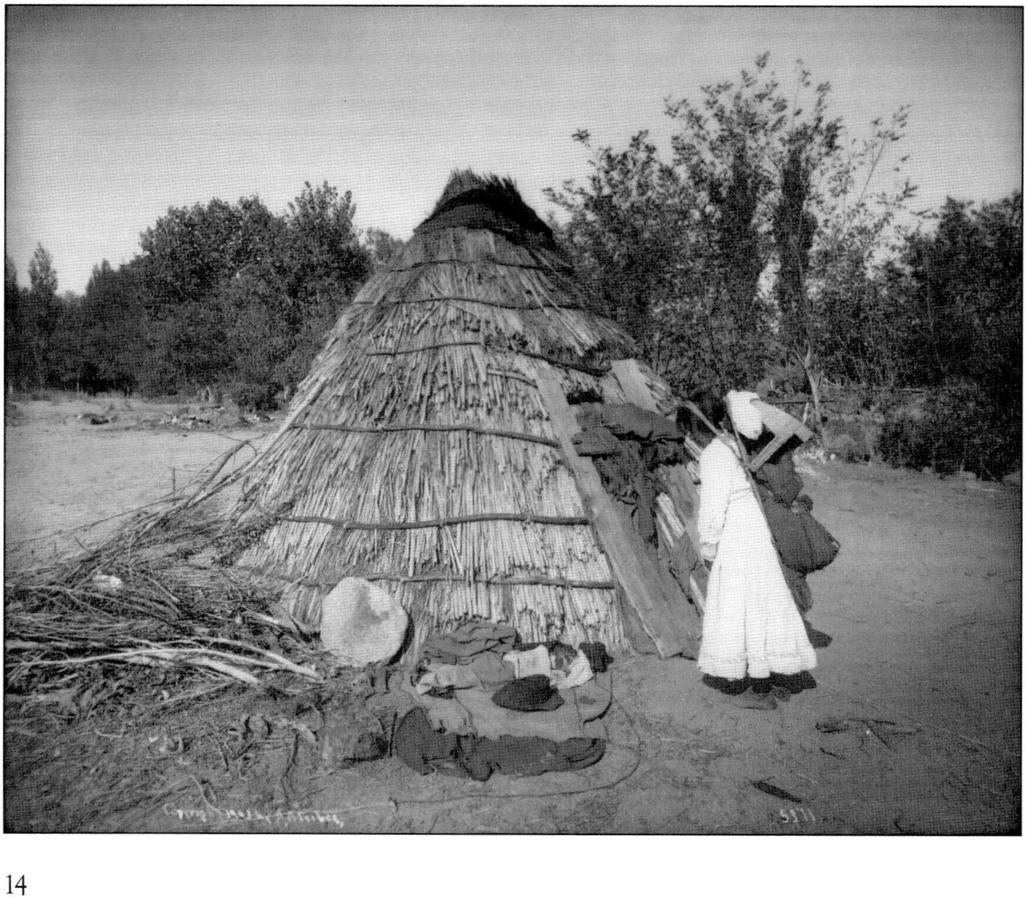

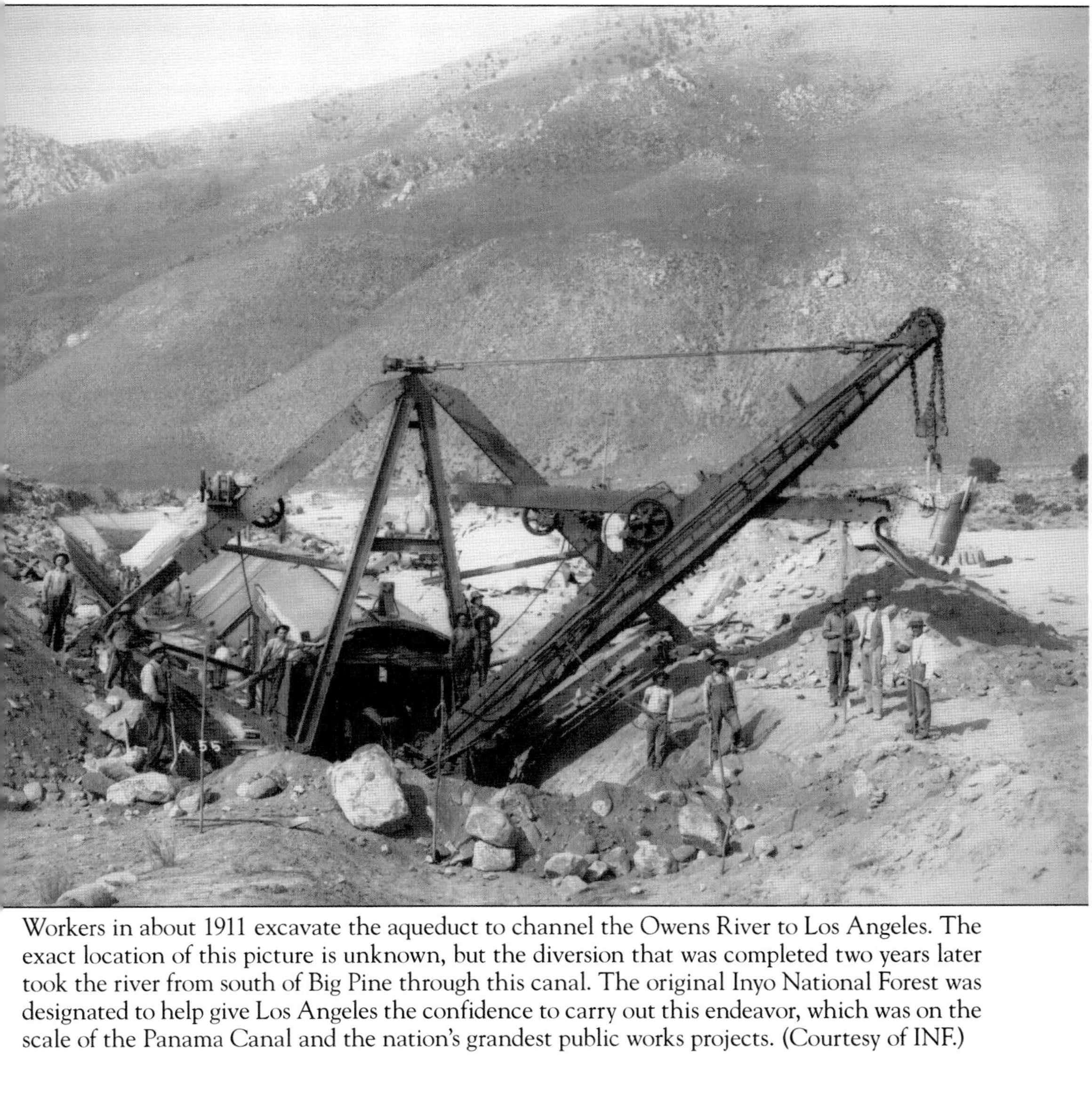

Workers in about 1911 excavate the aqueduct to channel the Owens River to Los Angeles. The exact location of this picture is unknown, but the diversion that was completed two years later took the river from south of Big Pine through this canal. The original Inyo National Forest was designated to help give Los Angeles the confidence to carry out this endeavor, which was on the scale of the Panama Canal and the nation's grandest public works projects. (Courtesy of INF.)

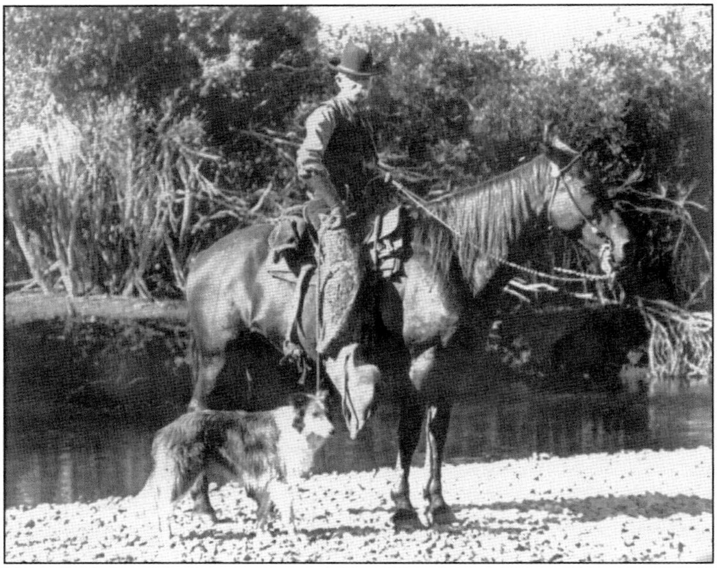

This wrangler named Wells is said to have been the first "eastside" federal ranger, representing the Sierra National Forest. He built the area's first ranger station about a dozen miles northwest of Bishop at the base of the Wheeler Crest, a location now known as Wells Meadow. This photograph is dated 1905, three years before the Sierra eastside was transferred to the Inyo National Forest. (Courtesy of INF.)

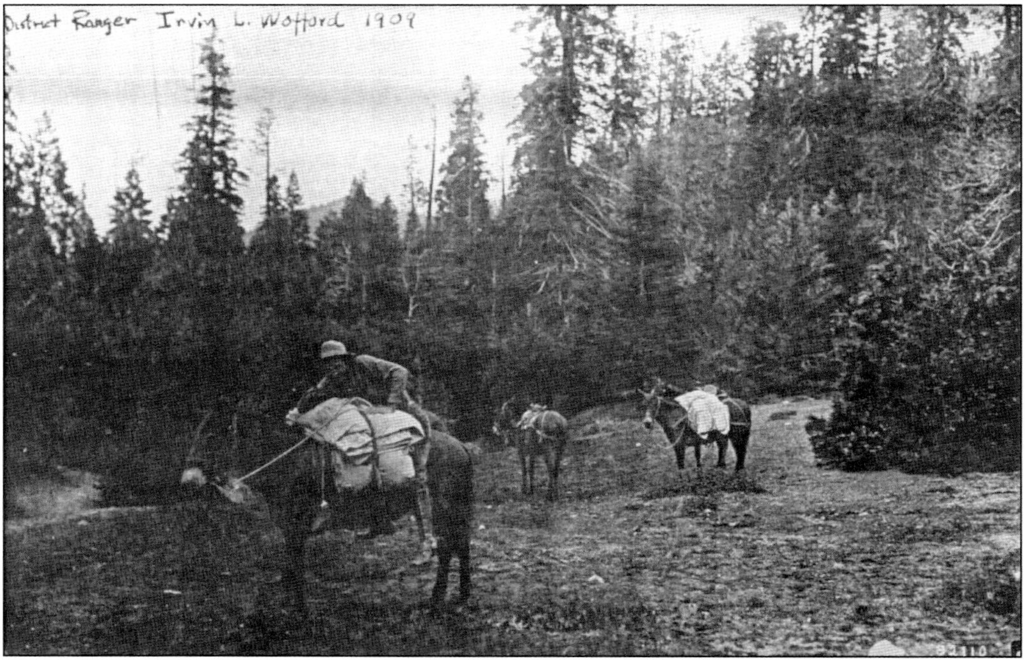

This is "District Ranger Irving L. Wofford, 1908." He likely was working on the Kern Plateau on land that in 1920 became part of the Inyo. Any ranger in the Sierra then had to be their own packer, farrier, cook, trailblazer, and all-around wrangler as well as regulator of the permitted activities on the forest. (Courtesy of INF.)

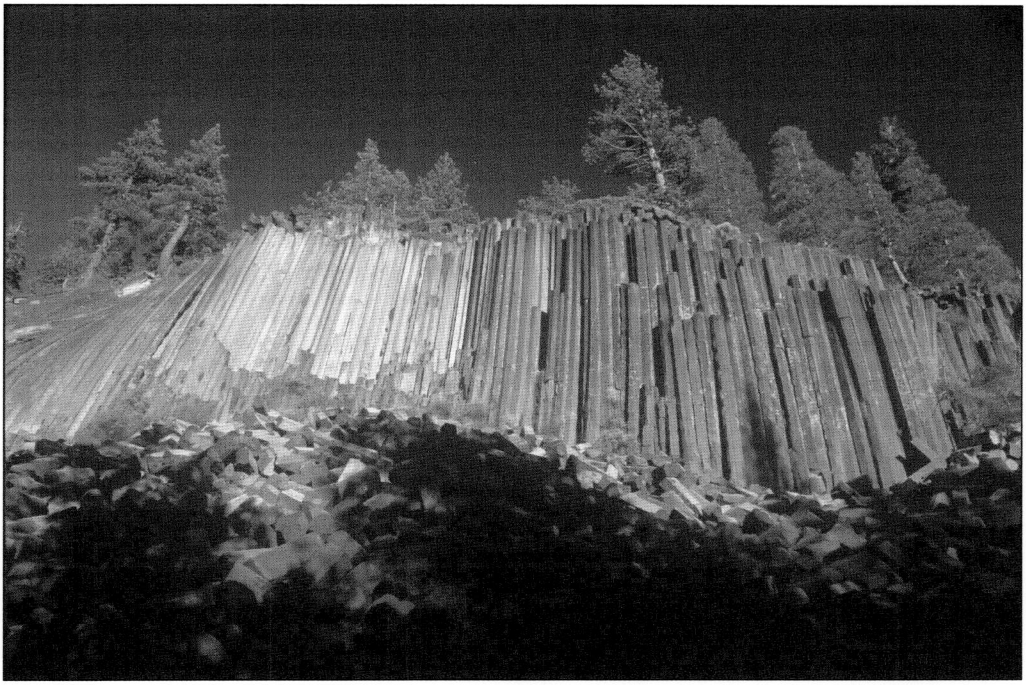

Devil's Postpile and Rainbow Falls are on the upper San Joaquin River, west of the Sierra Crest, but accessed through the Inyo. These columnar volcanic cliffs were actually part of the original Yosemite National Park, but in 1905, mining interests successfully lobbied to access the area with a transfer to the Sierra National Forest. Then, they submitted a plan to blow up the basalt columns and use them as fill for a hydroelectric dam. The Forest Service called this a "wanton destruction of scenery" but had no legal authority to stop the plan. They notified President Taft, and in 1911 he designated the canyon as Devil's Postpile National Monument. Eventually, the monument was transferred to the National Park Service. Now, it is an enclave within the Inyo's Mammoth Ranger District and a very popular destination for visitors. (Courtesy of AS.)

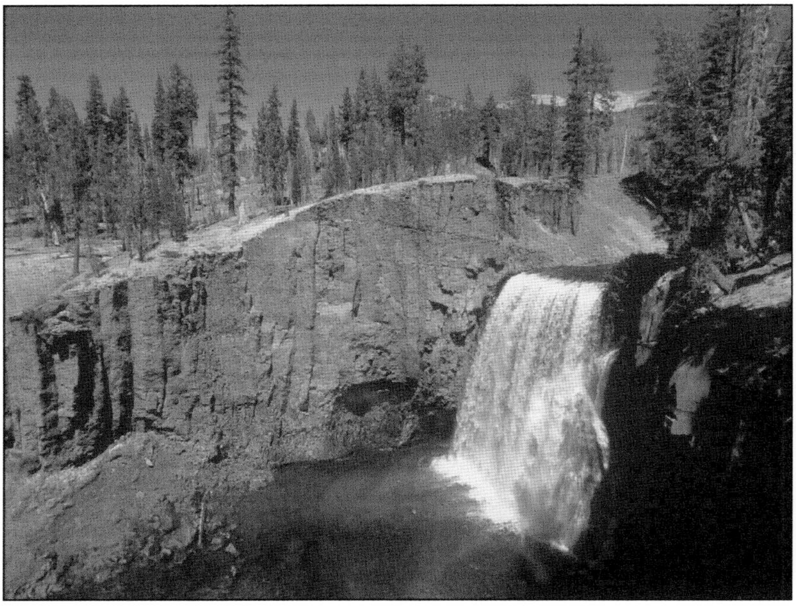

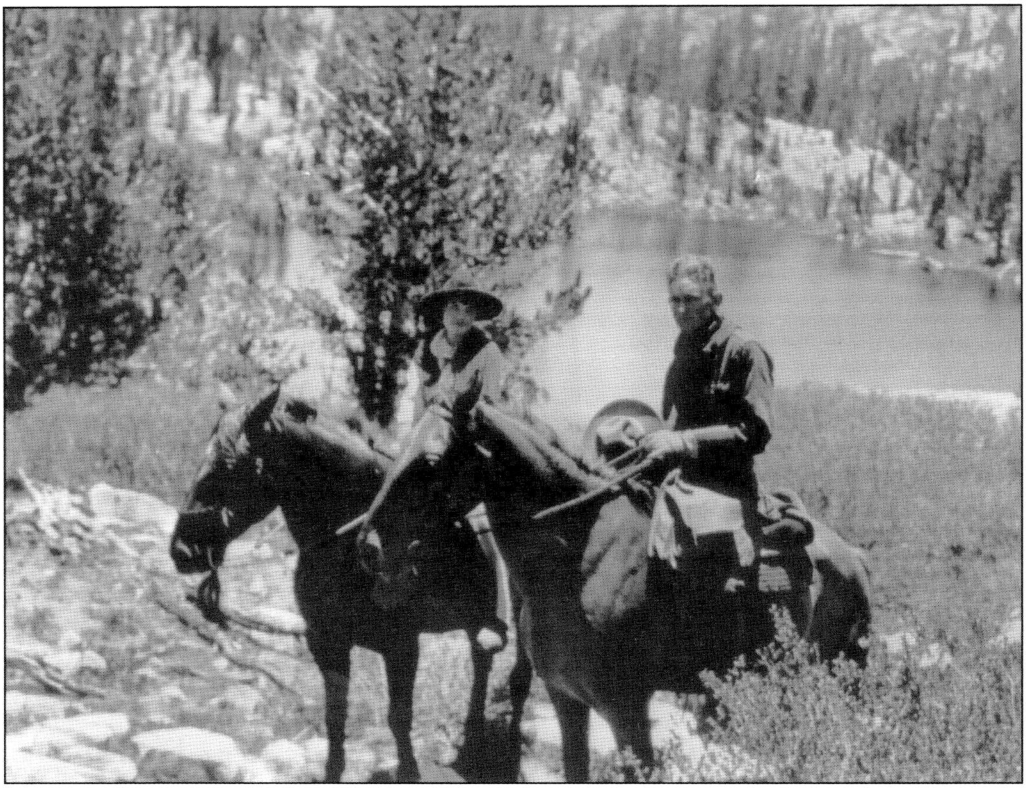

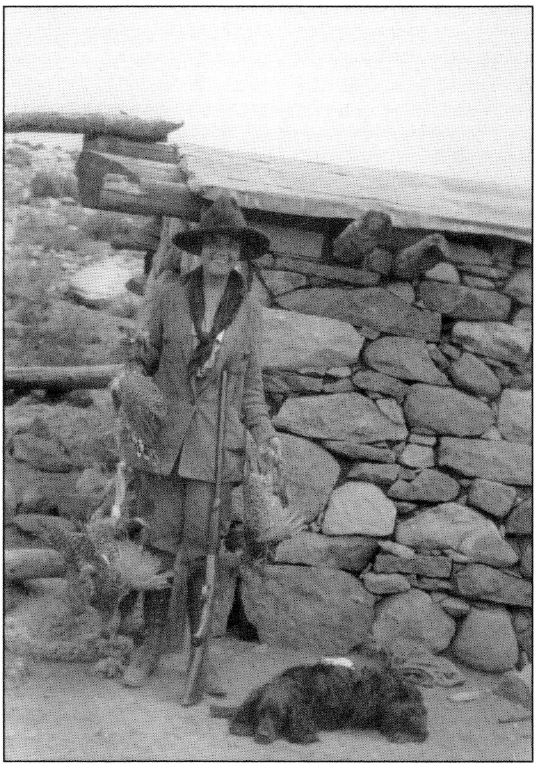

Ranger Lawson Brainerd and his wife, Helen, pause near Summit Lake in the Big Pine drainage. Both of them fell in love with the Inyo, and Helen usually accompanied her husband on his field assignments. One of the lakes in the Big Pine south fork drainage is named (albeit misspelled) for him. Though he worked for only six years on the Inyo, his can-do attitude and lighthearted writing make him a legend for staff today. (Courtesy of INF.)

Helen Brainerd holds her limit of four sage grouse she hunted at Coyote Flat, a high ridge above Bishop. The area was an important sheep-grazing allotment, and when her husband, Lawson, went to patrol and camp there, she shared in the adventure, too. (Courtesy of INF.)

This is Wes Hotteling, Big Pine district ranger, in 1922. In the green wool uniform that rangers wore while working in town, he heads to the post office with the day's mail. (Courtesy of INF.)

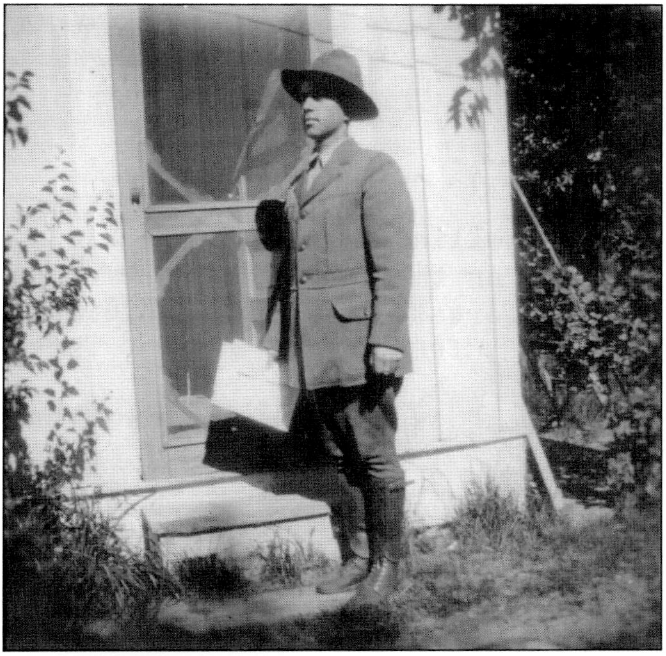

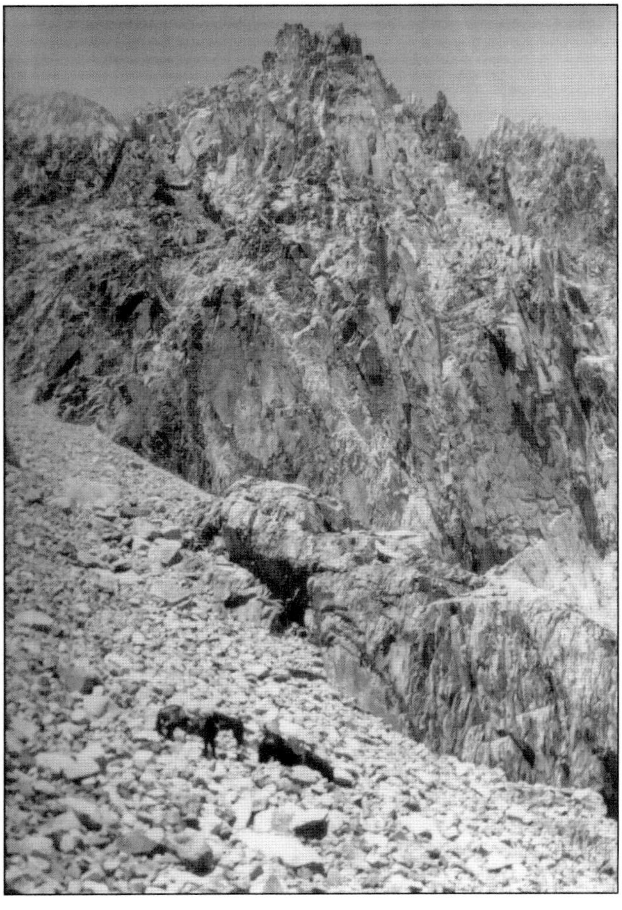

Lawson Brainerd took this photograph in 1924 or 1925 while he was forging a rough horse trail over 11,300-foot-high Taboose Pass, located southwest of Big Pine. "I tried to locate the best route and with some money snitched from that allocated to other trail projects worked out a path that was not such a horse killer," said Brainerd. (Courtesy of INF.)

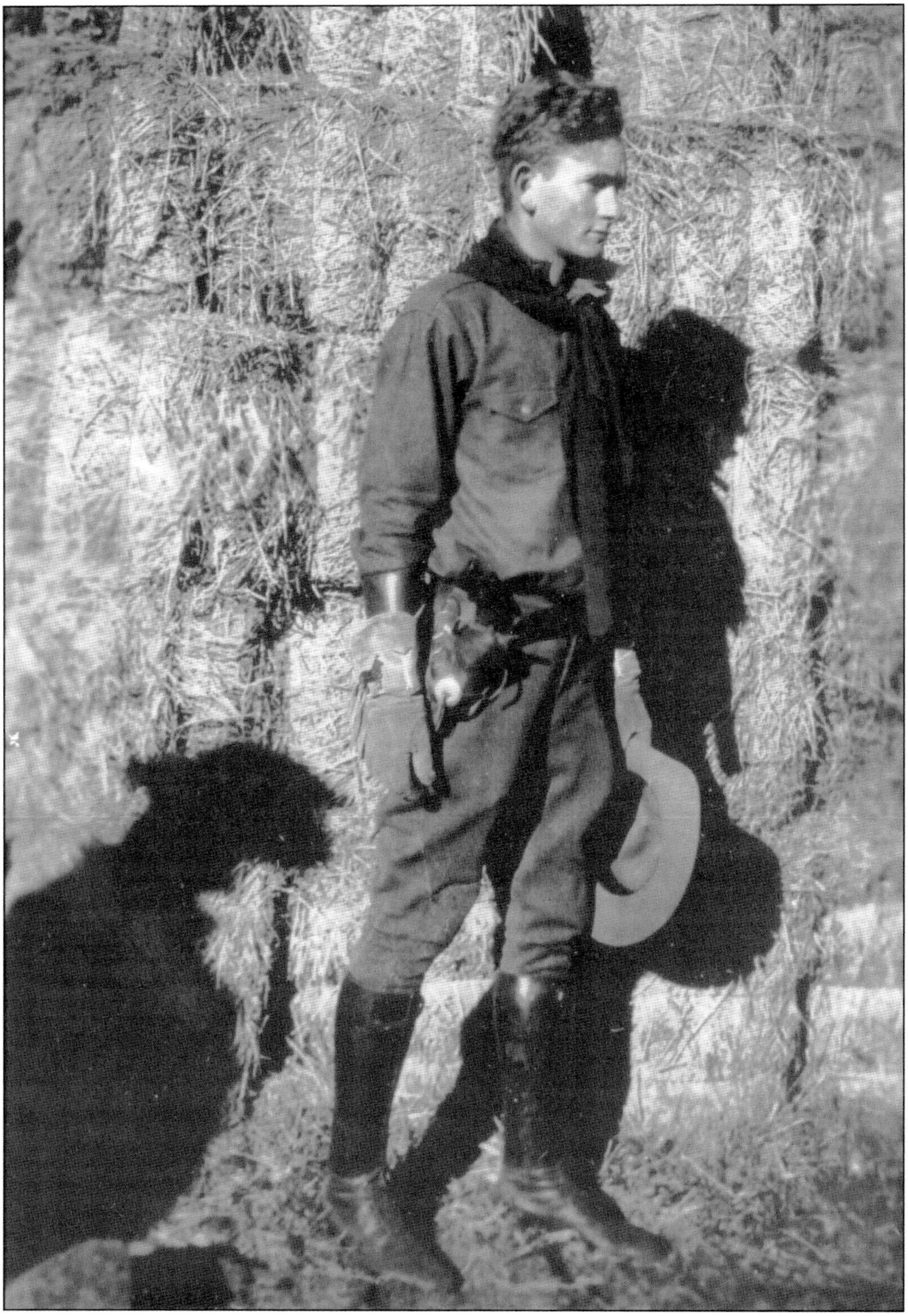

Lawson Brainerd is shown here at the Inyo's hay stockpile and corral. He later wrote, "If St. Peter, in a careless moment, allows me to pass through those Pearly Gates, my hope is that it will be to again be District Ranger on the White Mountain-Big Pine District, with Roy Boothe as my supervisor. With old Buck prancing under me and Helen beside me, Fossil Face, Winneduma, and Cactus trailing behind us, mountains around us, the fresh, cold wind in our faces—that will be our Nirvana, Kismet." (Courtesy of INF.)

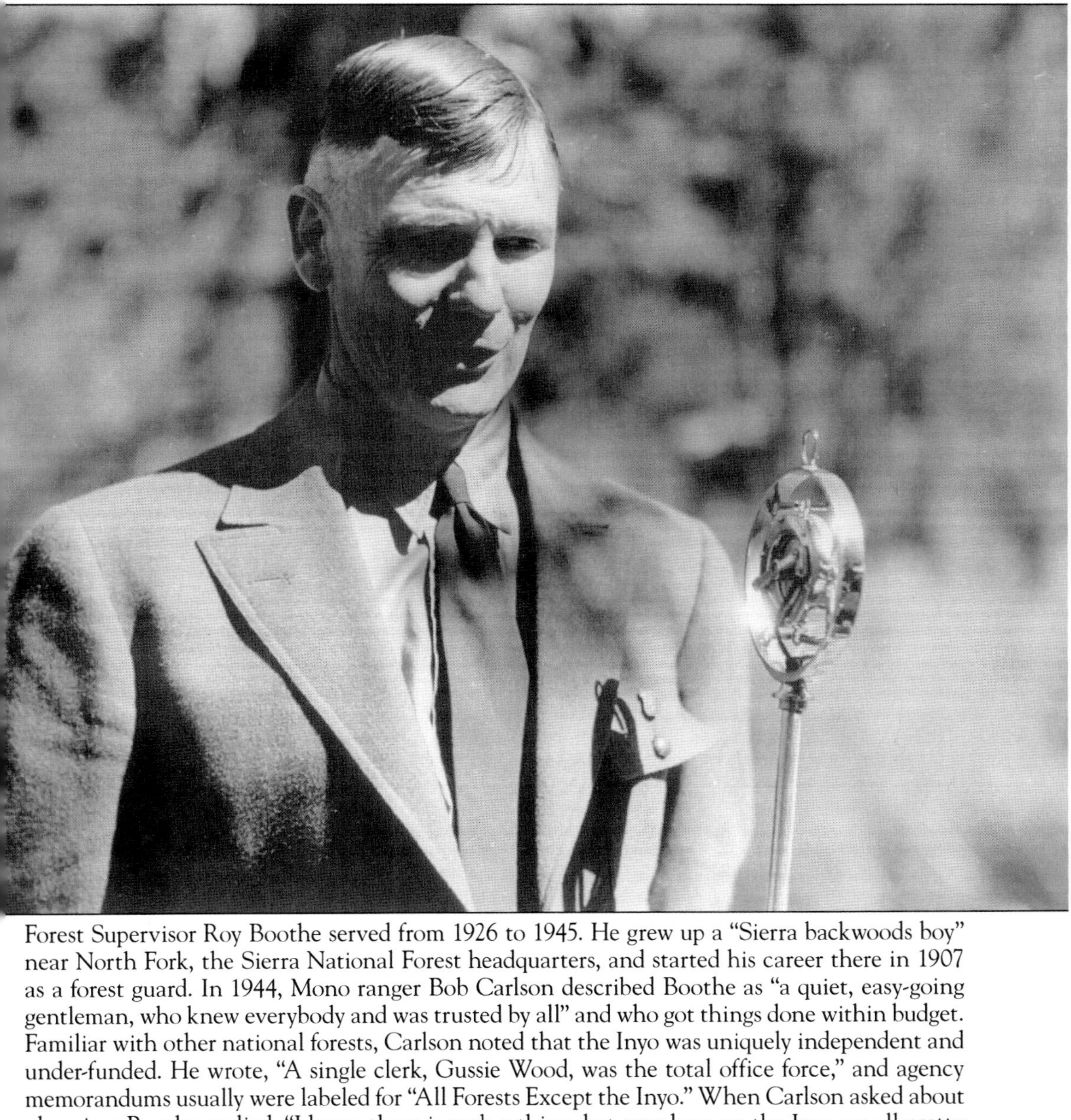

Forest Supervisor Roy Boothe served from 1926 to 1945. He grew up a "Sierra backwoods boy" near North Fork, the Sierra National Forest headquarters, and started his career there in 1907 as a forest guard. In 1944, Mono ranger Bob Carlson described Boothe as "a quiet, easy-going gentleman, who knew everybody and was trusted by all" and who got things done within budget. Familiar with other national forests, Carlson noted that the Inyo was uniquely independent and under-funded. He wrote, "A single clerk, Gussie Wood, was the total office force," and agency memorandums usually were labeled for "All Forests Except the Inyo." When Carlson asked about planning, Boothe replied, "I know there is such a thing, but over here on the Inyo we all pretty much know what we have to do and go out and do it. We don't see the need for a bunch of paperwork, so there is no analysis on the Inyo." Boothe is shown here on May 31, 1936, speaking at the inauguration of the new road to Whitney Portal. This was an outdoor banquet with media and dignitaries, including Governor Merriam. (Courtesy of INF.)

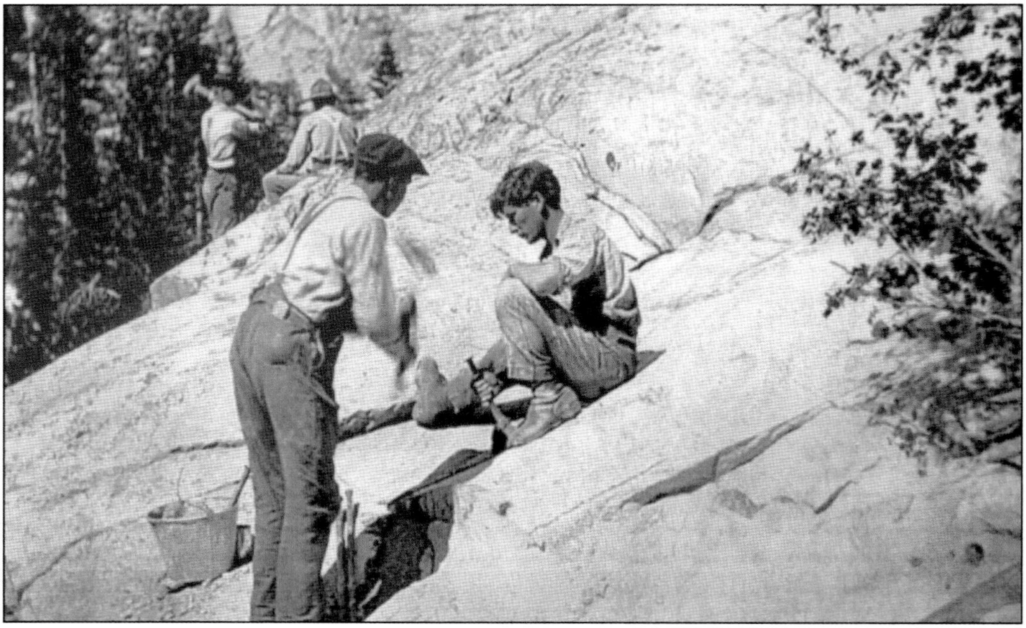

Roy Boothe had overseen the 1918 start of construction of the John Muir Trail when he was a ranger on the Sierra National Forest. When he became the Inyo supervisor, he took trips over Bishop Pass to continue the work. The "Pathway in the Sky" would run from Mount Whitney to Yosemite, through 215 miles of high country grandeur. In the first picture, pairs of workers in LeConte Canyon are using hand-held drills and sledgehammers to drill into solid granite. Dynamite was stuffed into the holes to blast the rock away. Below, a worker uses a field forge and anvil to make and maintain tools and horseshoes. (Roy Boothe photographs, both images courtesy of Helen Dixon.)

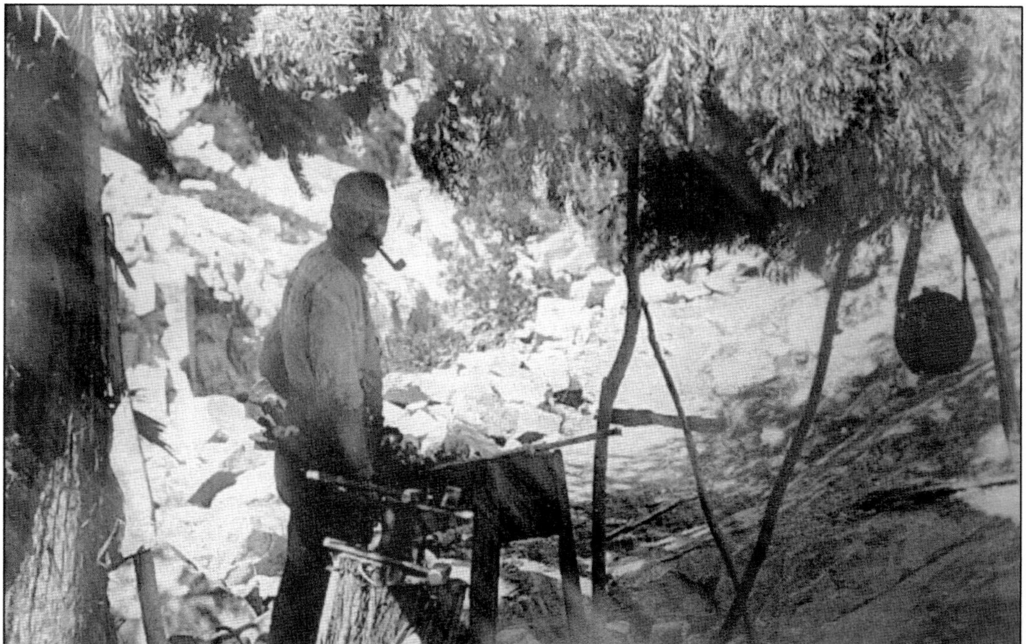

Two

Harvesting the Stuff of Life

Natural Resources

Congress mandated the Forest Service to manage the public lands for a sustainable supply of resources, particularly wood but also forage and water, too.

The Inyo was endowed with some areas of harvestable timber, notably the Jeffrey pine and red fir forests in the snowy lee of Mammoth Pass as well as the white fir forests above Owens Lake. For sheep and cattle, there was adequate forage in the high country of most canyons. But the most vital commodity here was water. In the arid West, water tumbling out of the mountains is precious not only for drinking and farming but also for the generation of electricity. Early engineers salivated over the tremendous "head" of the Sierra's crashing streams, and some of the nation's earliest and most remote mountain dams and hydroelectric facilities were developed here.

Mining continued on the Inyo throughout much of the 20th century at various scales on various mountainsides, but the trend shifted from boomtown prospectors to larger, more industrial operations, often excavating for strategic minerals.

Early travel, even along the mountain fronts, was a substantial challenge, and then getting up into the mountains was essentially an expedition. With time came, of course, a steady progression of route development, from trailblazing that tended to follow the Paiutes' natural paths to trail and road construction by miners, packers, and, occasionally, national forest rangers. Eventually, highway engineering climbed up several of the canyons, making a summer trip up to some of the lakes a matter of casual motoring.

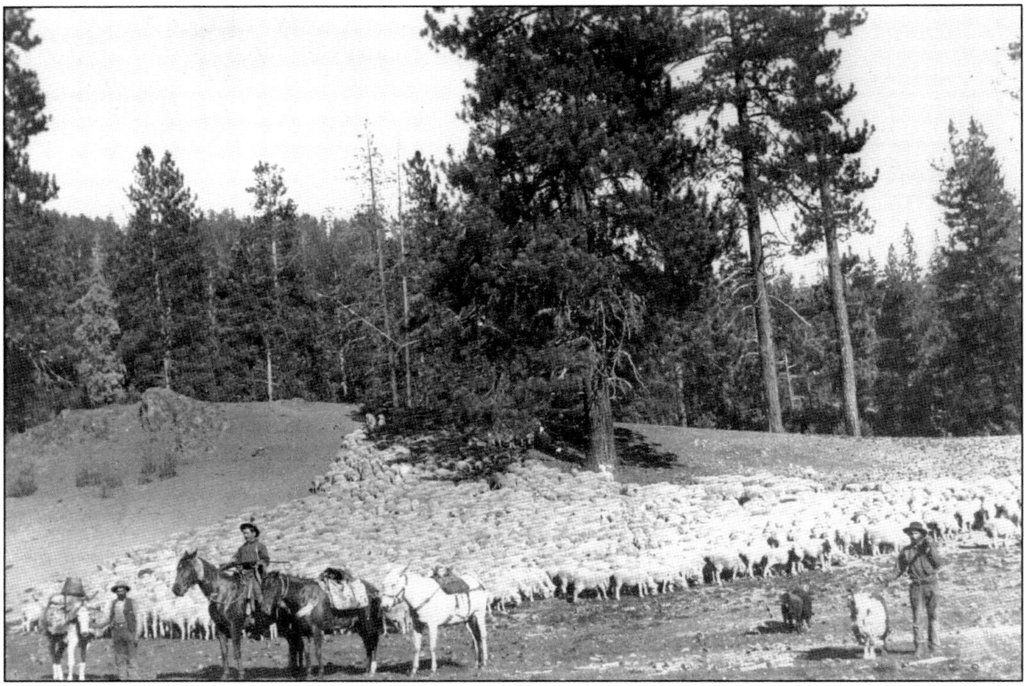

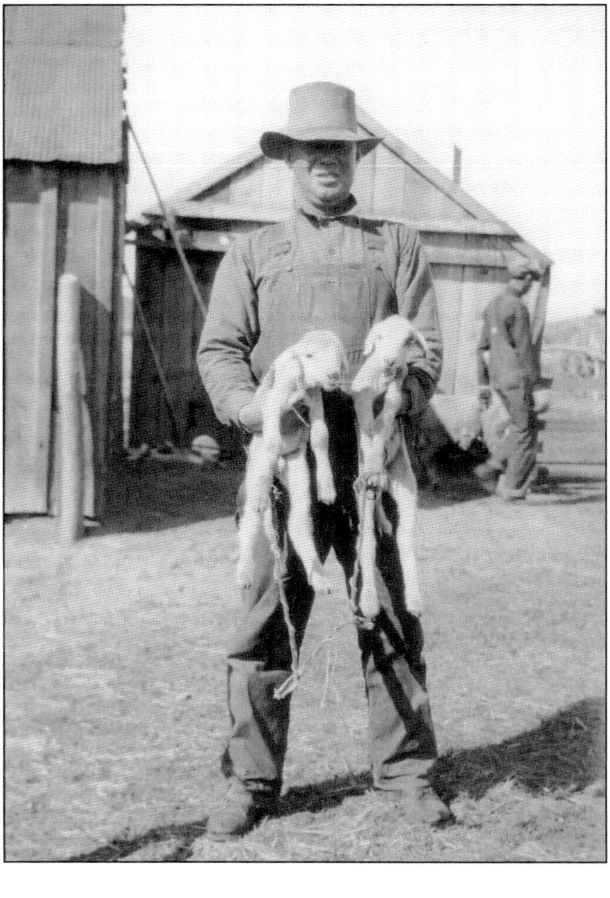

Sheep forage was one of the primary resources on the Inyo, and monitoring sheep allotments took the bulk of the staff's time. Turning the forage into wool and mutton was an important and often lucrative industry, but operations were competitive and infamously prone to overgrazing. John Muir, who worked as a shepherd in his first summer in the Sierra, famously called sheep "hoofed locusts." It is also thought that the pneumonia bacteria that domestic sheep carry caused the near decimation of the native Sierra bighorn. Above is an 1897 photograph of Pete Giraud's sheep at Templeton Meadow on the Kern Plateau. At left is George Baker, whose family operated a sheep ranch near Big Pine for decades. The Baker family members sent their animals up into the Inyo during the summer until around World War II. (Courtesy of INF.)

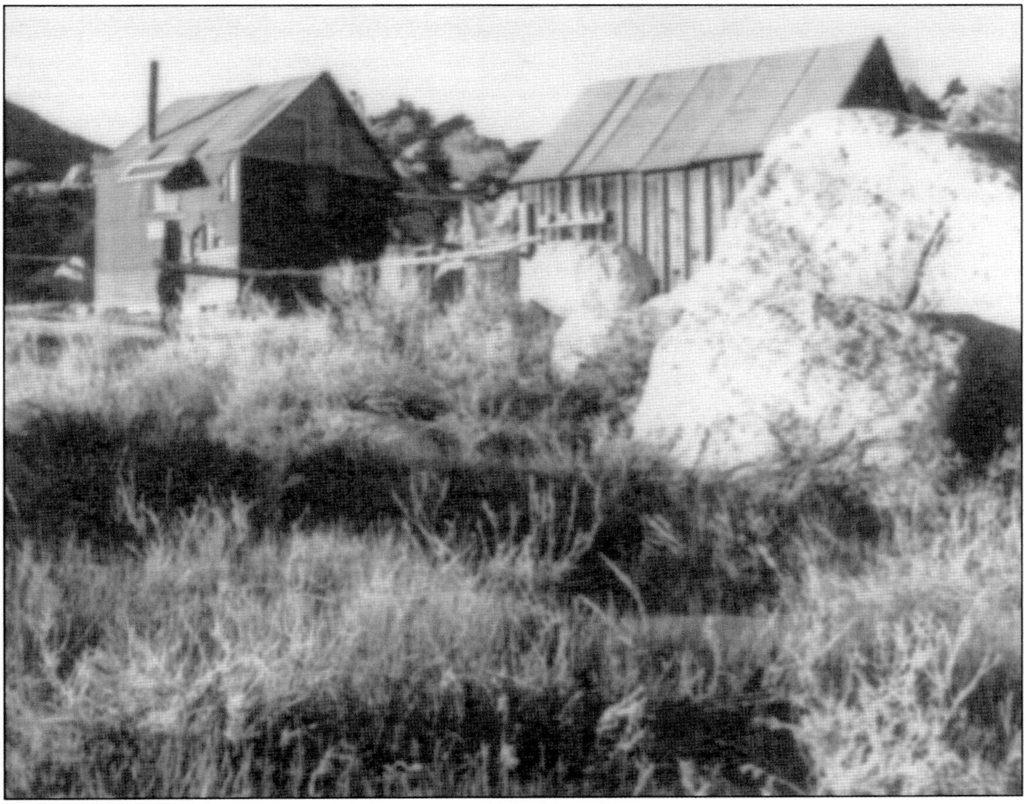

In the arid White Mountains, the Inyo allowed grazing through the World War II era. Especially when rangers were few, it was common practice for a permittee to set out more sheep and use more territory than they were allotted. Ranger Lawson Brainerd said, "Raising sheep was like gambling. If everyone played square . . . one might come out ahead, but, if everyone but you cheated, you lost." Brainerd took the picture below, of sheep in the White Mountains. He patrolled out of this station at Crooked Creek (above). This site, at 10,200 feet, now contains a research facility for the White Mountain Research Station. (Courtesy of INF.)

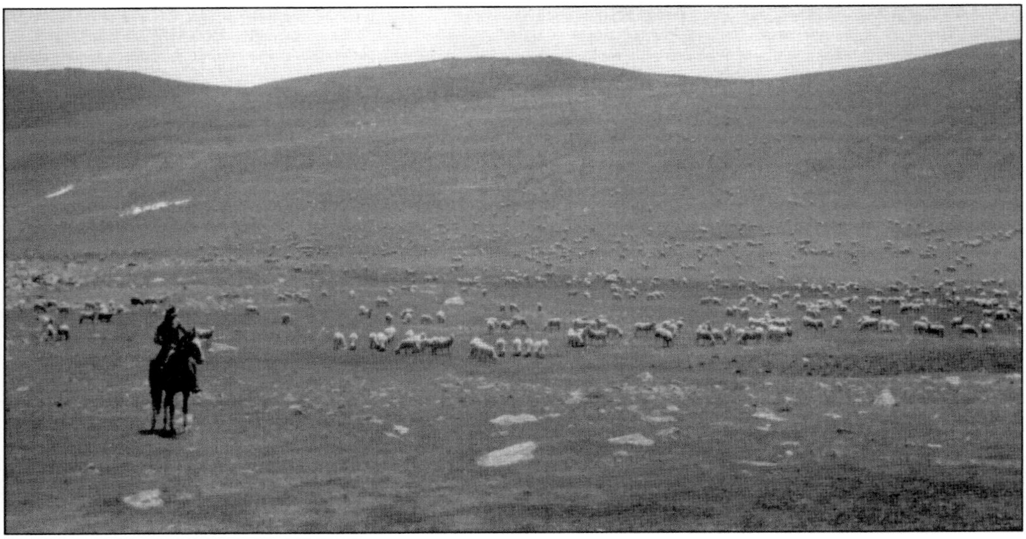

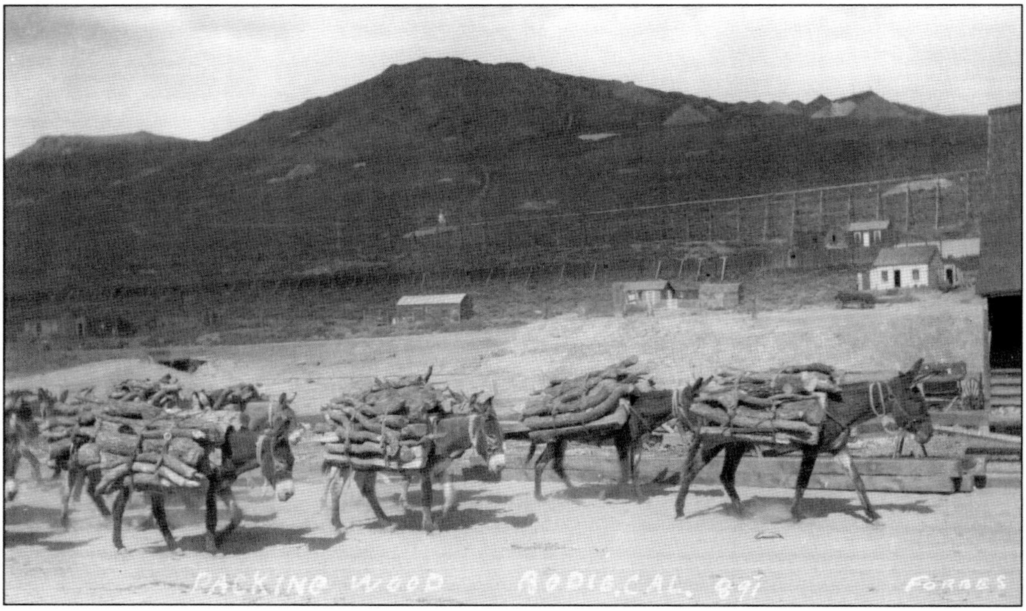

The forests on the eastern slopes of the Sierra were not tremendous, but they were nevertheless the closest wood for building and fueling the region's mining operations. Here, a train of donkeys carries firewood from the Sierra to Bodie, the region's largest mining town. Other operations harvested trees from most every major canyon in the Sierra. (Courtesy of ECM.)

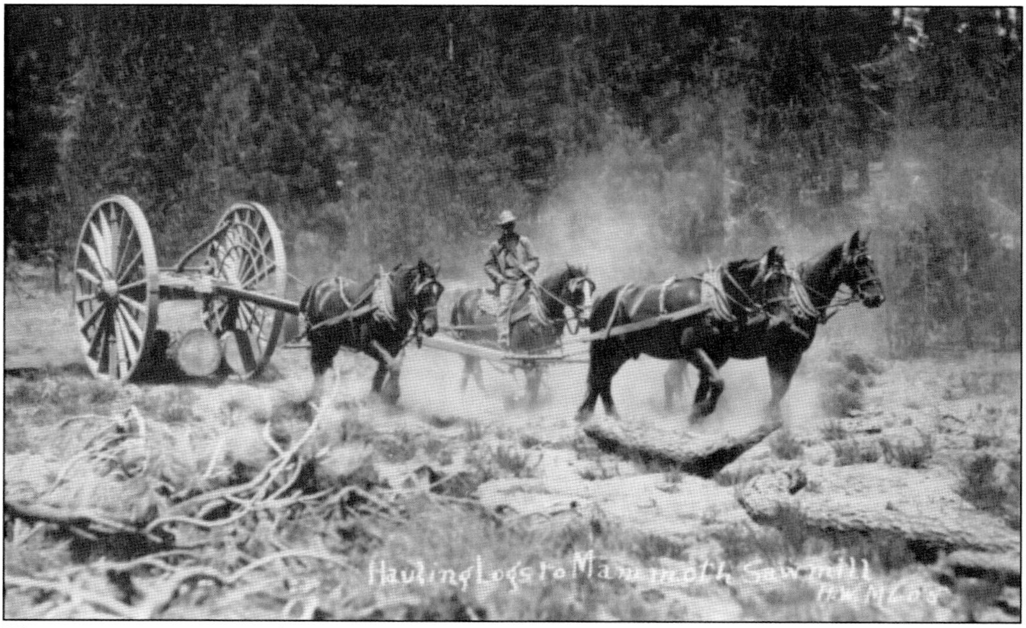

The snowy region around Mammoth was and still remains the most forested part of the Inyo. Tall Jeffrey pines and red firs were hauled to a sawmill operated by Arthur Hess, near what is now Shady Rest Park. Deep, sandy, volcanic soil made for difficult hauling, and the specially designed under-axle carriage with large, spoked wheels helped. More recent logging operations have focused on hauling in winter, over snow. (Courtesy of ECM.)

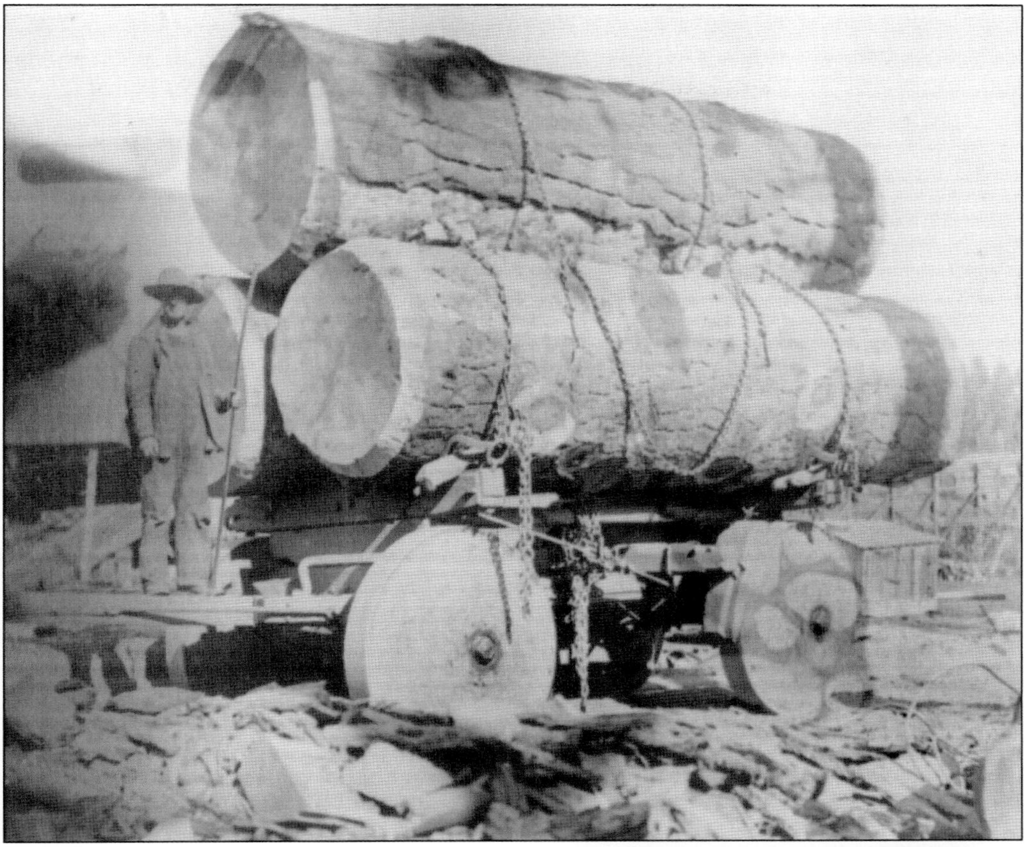

South of Mount Langley, the Sierra breaks into a relatively gentle terrain that supported stands of impressive white fir and other conifers. In the 1870s, Cerro Gordo mining operations in the Inyo Mountains had used up the piñon pines there, so Sherman Stevens endeavored to bring them wood from the Sierra. Access up and down the Sierra front was extremely difficult, so his company built a sawmill at nearly 10,000 feet and a flume to send logs, lumber, and even men careening down into the Owens Valley, about 6,000 feet below. Above is an ox wagon heavily loaded with white fir, and at right is a section of the flume. When the Cerro Gordo Mine played out, Stevens's operations sent wood south to the next boomtown, Darwin. (Above, courtesy of INF; right, courtesy of ECM.)

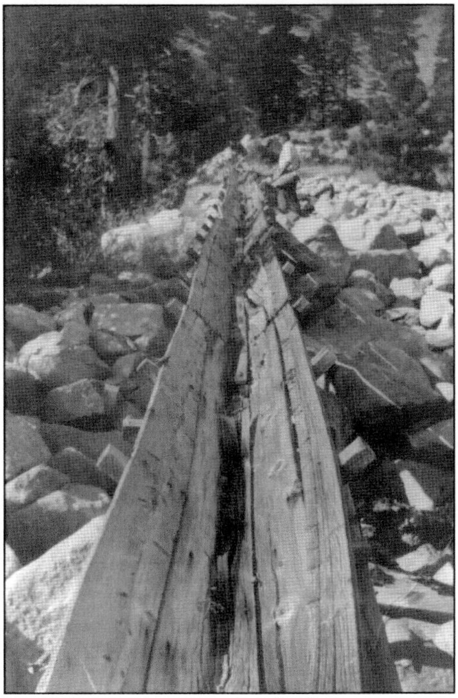

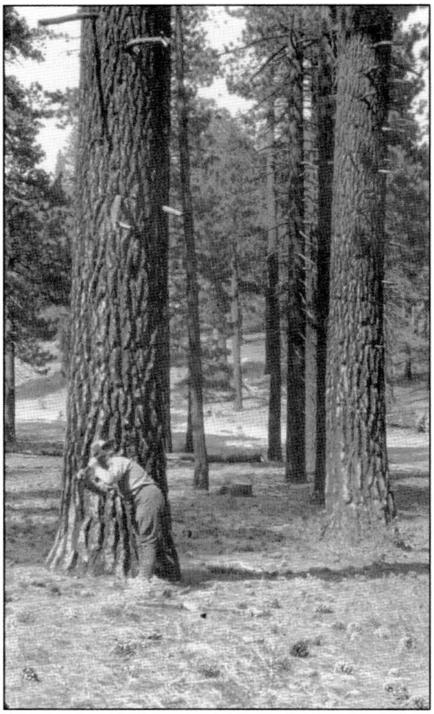

An Inyo ranger marks Jeffrey pines for harvest for the Hess logging operation in 1924. Note the character of the stand, with tall, mature trees and little underbrush. This reflects a condition of relatively frequent, low-intensity fires. In subsequent decades, fire suppression combined with harvesting of the larger trees would result in these forests growing denser and more fire-prone with young, less-healthy trees. (Courtesy of INF.)

Forest Supervisor Joe Radel (left) inspects logs at the Inyo Lumber and Milling Company. The facility was located two miles west of Bishop. Behind them are the mill and a "teepee burner," where bark and other debris were disposed of. Mill owner Lou Moffat was affectionately known as "Low-Profit Moffat" because his company struggled with long distances to markets and a slim timber base. Most of the logs were milled into box lumber. Within a few years after this picture was taken in the mid-1950s, the operation closed down, and Inyo logs were hauled to mills farther away. (Courtesy of INF.)

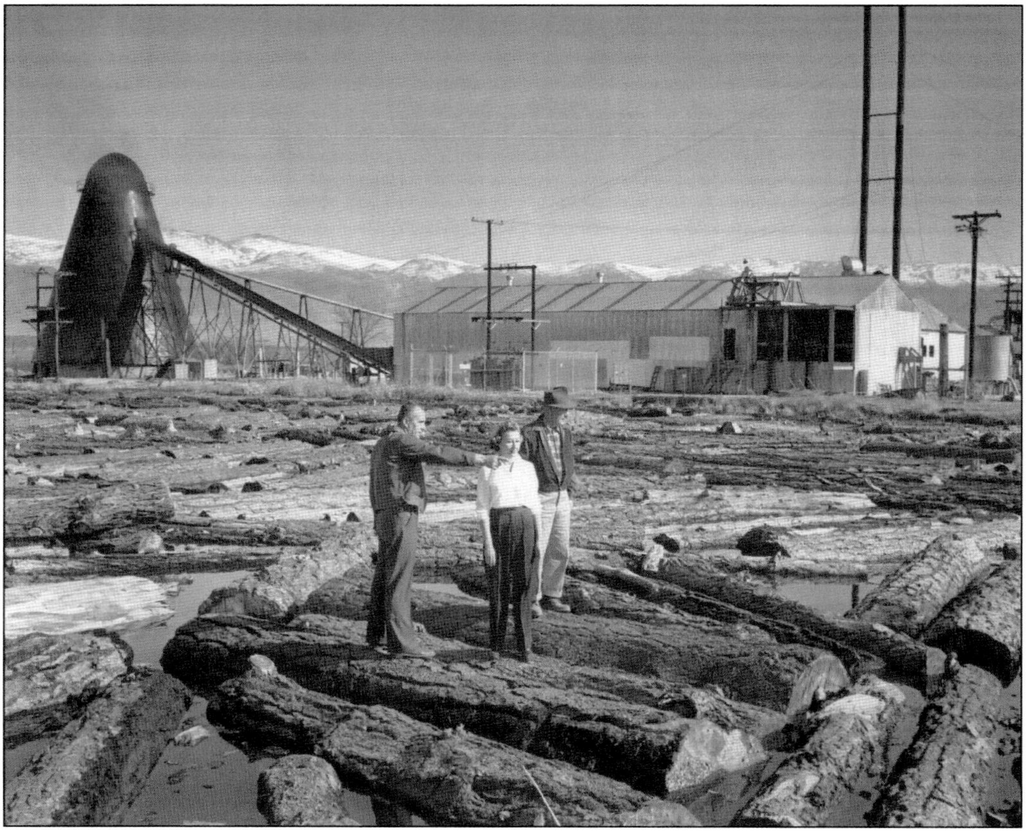

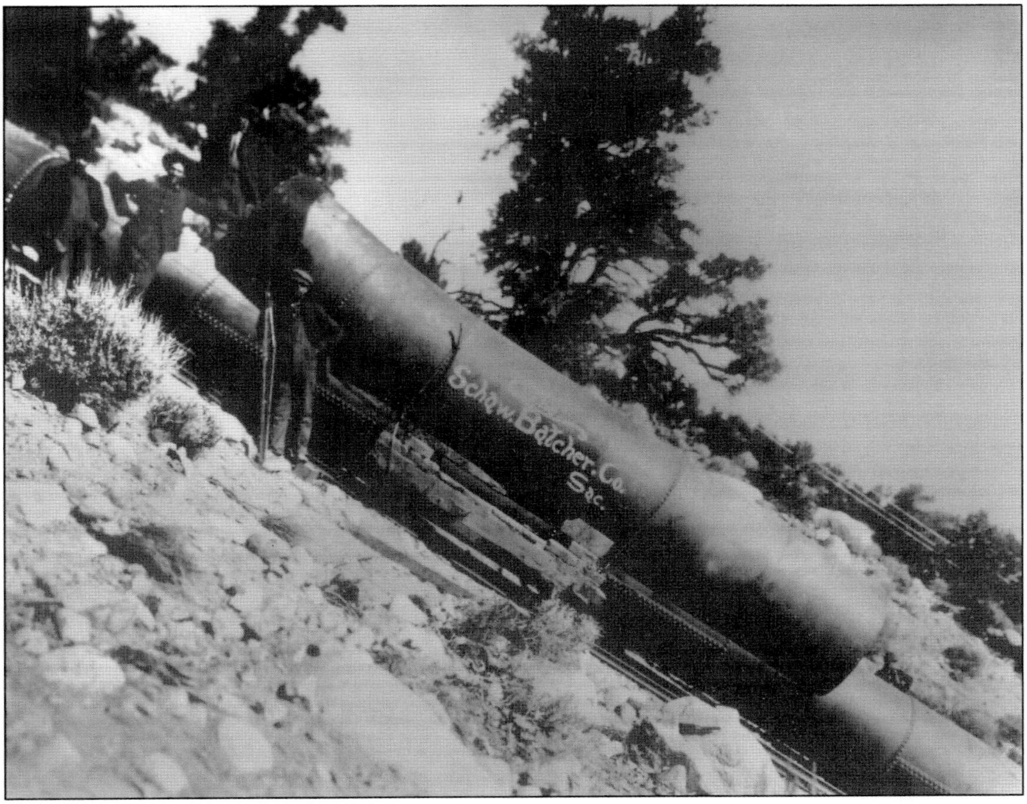

In 1905, the Nevada Power, Mining & Milling Company embarked on one of the world's most ambitious hydroelectric schemes to that time. The gold and silver booms in Tonopah and Goldfield, Nevada, needed power, and Bishop Creek was the nearest tumbling water. By 1919, they built two reservoirs, South Lake and Lake Sabrina, and harnessed 3,590 feet of "head" (elevation drop) to send electricity 113 miles over the White Mountains, along probably the world's longest transmission line. They built five plants in series along the creek, generating an average of 160 million kilowatt hours per year. Above, workers pause in the laying of penstock pipe. When the mining boom played out, they strung wires the unimagined distance of 240 miles, to sell power in San Bernardino. The picture below shows the shoring up of hasty initial construction of Sabrina's dam to withstand the coming winter of 1908. All the Bishop Creek facilities are still operating today under Southern California Edison. (Above, courtesy of INF; below, courtesy of SCE.)

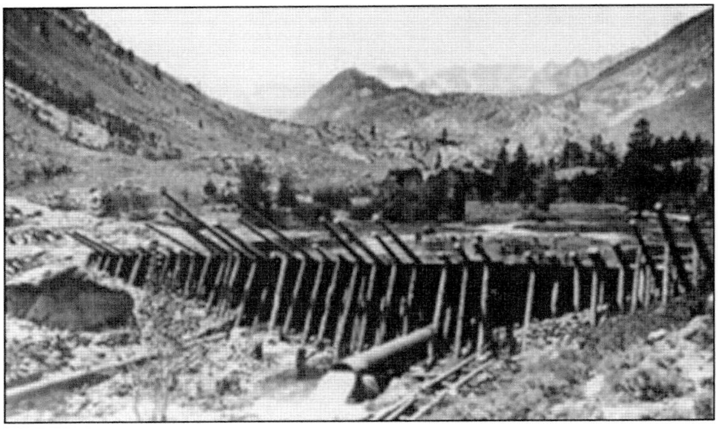

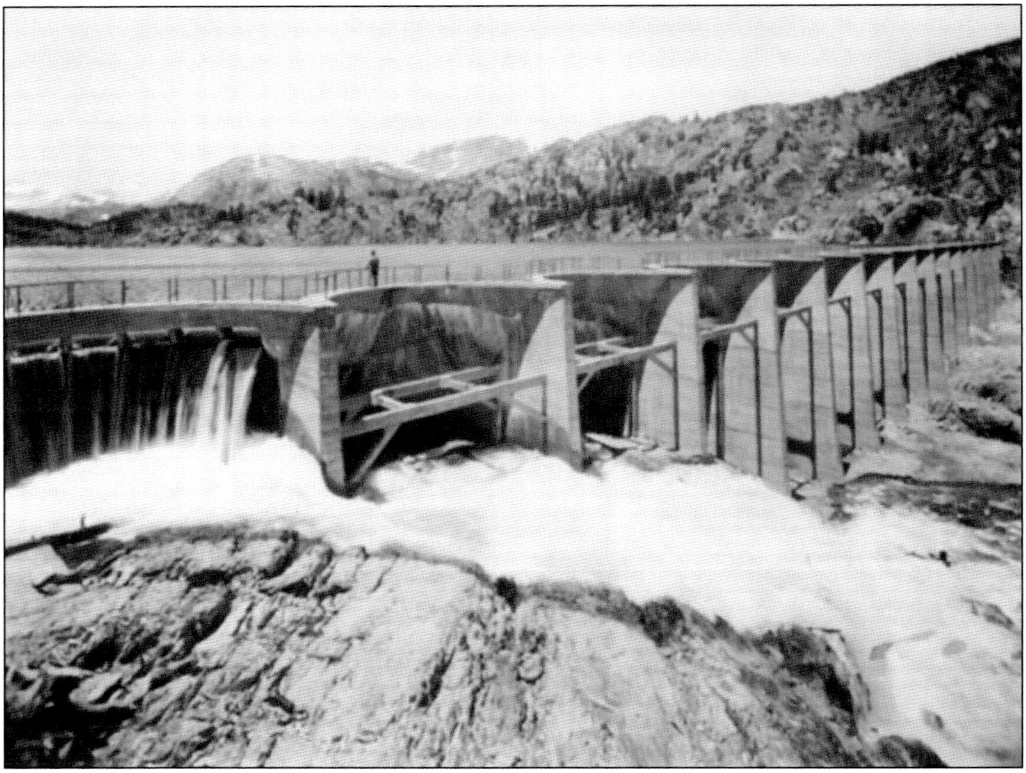

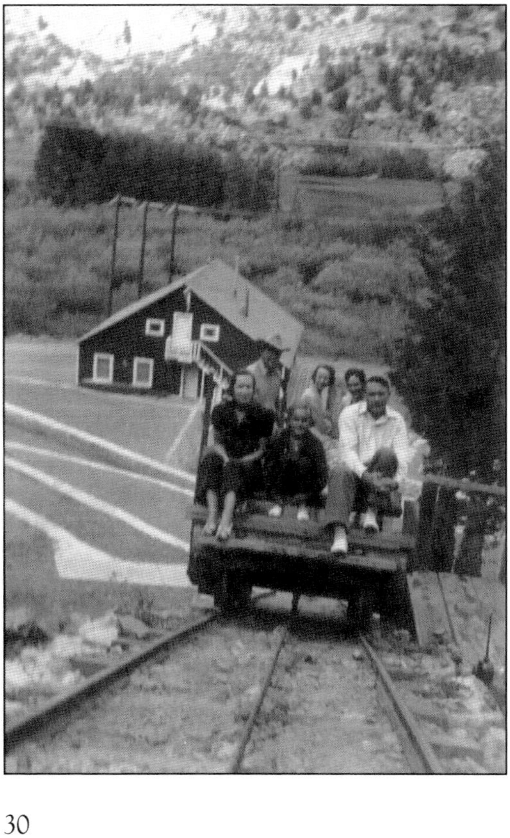

To harness the power at Rush Creek (in the Mono Basin), the Pacific Power Corporation built this multiple arch dam at 8,500 feet, enlarging natural Agnew Lake. A penstock from there down to a power plant near Silver Lake delivered about 1,300 feet of head. Under new management, two more arch dams were built upstream, enlarging Gem Lake and Rush Creek Lake (now known as Waugh Lake). (Courtesy of SCE.)

This "Incline Hoist" was built out of remnants from a Bodie tram to bring workers, concrete, and other materials up the steep slopes to build the dams on Rush Creek. A 75-horsepower, 440-volt motor pulled an 8,000-pound test rope to haul the tramcars up the tracks. At a rate of 320 feet of track per minute, it took just 15 minutes to get up the 1,329-foot climb. The tracks are still in place. (Courtesy of Vance B. Rhudy.)

Several companies raced to claim the hydroelectric potential at Rush and Lee Vining Creeks, and when the capitalist dust settled, the Nevada-California Power Company started developing the Lee Vining watershed in 1917. Just getting equipment and parts to the site stretched the company's resources. The Southern Sierras Power Company took over in 1922. Here, they haul with a bulldozer on snow. Finally, the Poole plant, named for the area's lead engineer since the Bishop Creek projects, was completed in 1924. (Courtesy of Mono Basin Historical Society.)

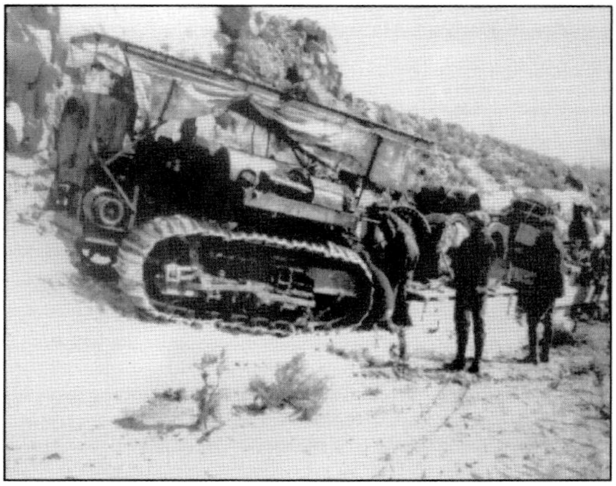

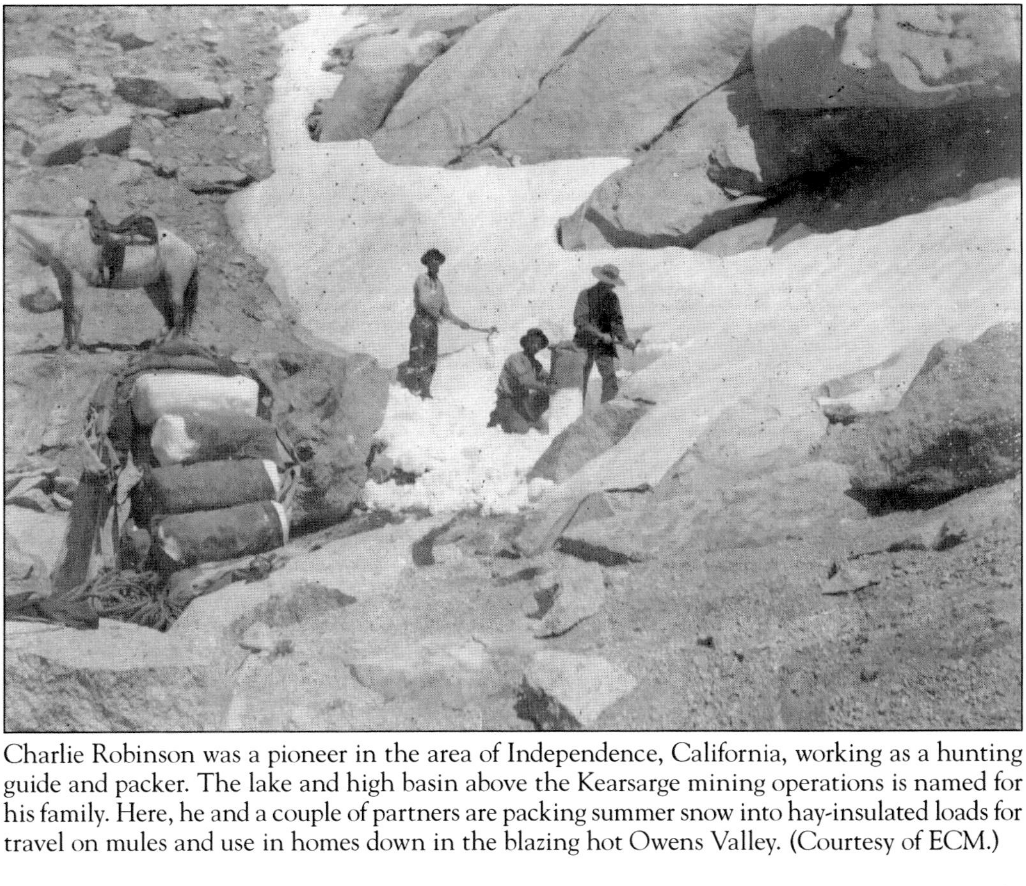

Charlie Robinson was a pioneer in the area of Independence, California, working as a hunting guide and packer. The lake and high basin above the Kearsarge mining operations is named for his family. Here, he and a couple of partners are packing summer snow into hay-insulated loads for travel on mules and use in homes down in the blazing hot Owens Valley. (Courtesy of ECM.)

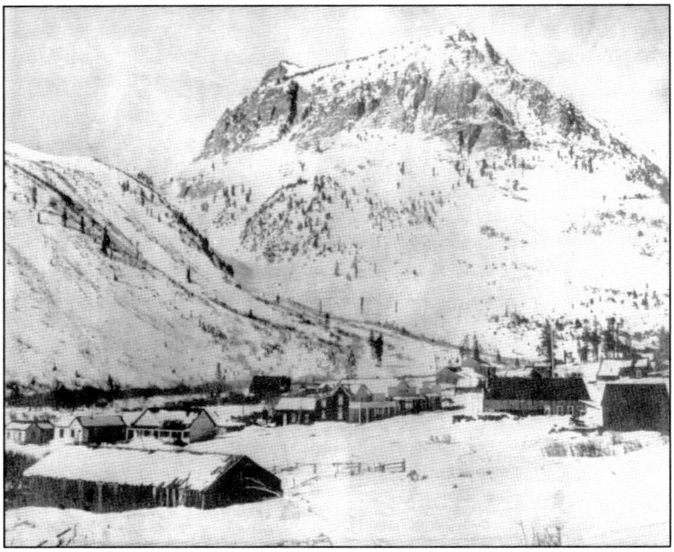

This is the mining town of Lundy, west of Mono Lake, in about 1907. The richest ore body here was found at over 11,000 feet in 1877. That mine was named May Lundy for the daughter of the man who had first run a sawmill here in 1864. The area was delineated the Homer District and produced some $3 million in gold. Up to 500 people lived here. Avalanches plagued the area, and a particularly large one killed several people and destroyed the Jordan power plant here in 1907. (Courtesy of INF.)

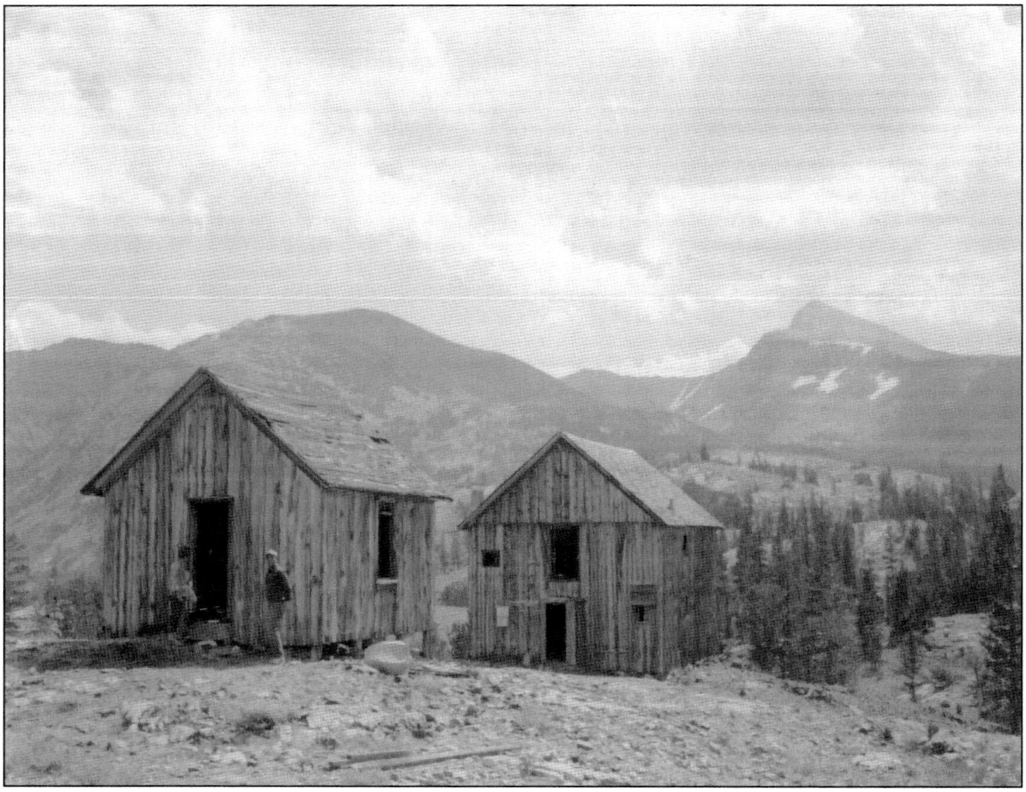

In 1878, a prospect above Lundy caused much excitement, and after many attempts to build access roads to the alpine location failed, eventually a company built a 56-mile-long wagon road from the Yosemite area. A town was built near the claims and named Bennettville for the company president, but excavations never found any valuable ore. In 1915, Stephen Mather bought the road, donated it to the National Park Service, and it became the Tioga Pass Road. This 1920s picture shows two remnants of Bennettville much as they remain today, thanks to National Forest maintenance of the historic structures. (Courtesy of INF.)

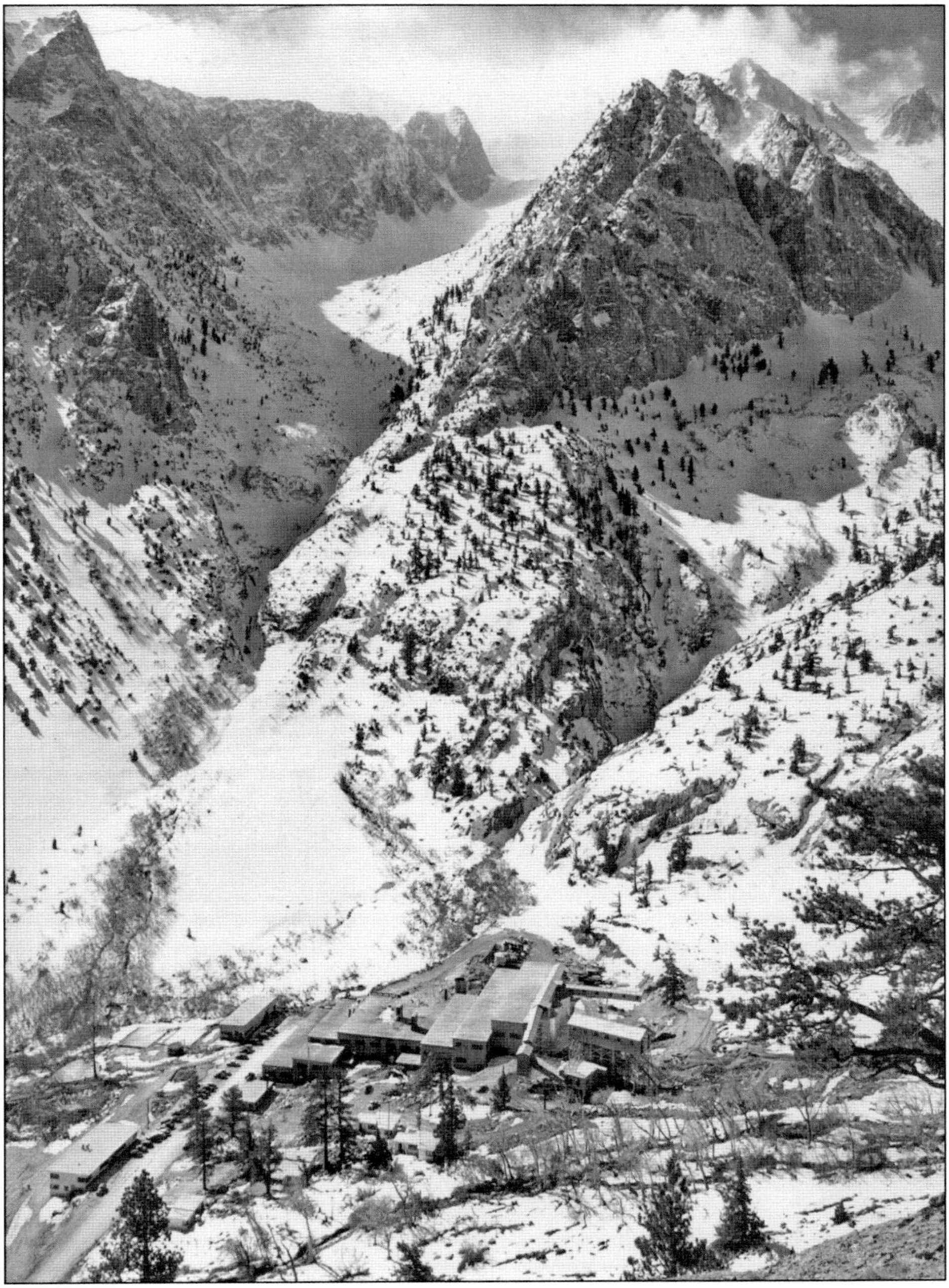

The US Vanadium Corporation built this facility at 8,000 feet in Pine Creek in 1941, and it quickly became the base for the largest tungsten operation in North America. Tungsten and vanadium are vital for hardening tool steel and, during both world wars, were strategic minerals. This mine operated for 53 years, excavating miles of tunnels within Mount Morgan (out of view to the right), collecting 37 percent of the United States' total output ever. This yielded enough tungsten to have made a 48-foot cube of the metal. At times, the release of nitrates and silt killed fish in Pine Creek. The mine was one of the principal employers in the region for decades. In 1990, operations wound down due to cheaper ore coming from China. (Curtis Phillips photograph, courtesy of John Boothe.)

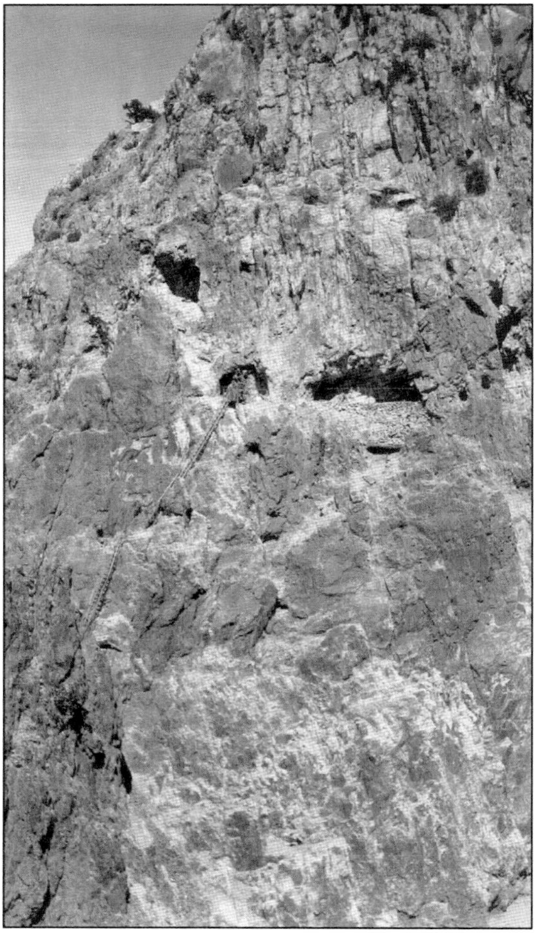

In 1914, these cliffs below White Mountain Peak were found to be rich in andalusite and sillimanite, alumino-silicate minerals used in the manufacture of ceramic spark plug insulation. When the United States entered World War I, the Champion Spark Plug Company quickly developed a road to the base of the mountains, a mule trail to this vertiginous ore body at 9,500 feet, and a high camp of buildings 1,000 feet below. By erecting scaffolding and blasting ledges, they began the difficult extraction of ore from the cliffs. To feed the miners and the mules, the company also established the White Mountain Ranch at the base of the alluvial fan below the mountains. The canyon was named for Champion's president, W.E. Jeffery, who is said to have enjoyed the remote and scenic outpost and kept the operation going longer than the mine's profitability (1942). The camp buildings are still maintained by devoted occasional visitors. (Courtesy of ECM.)

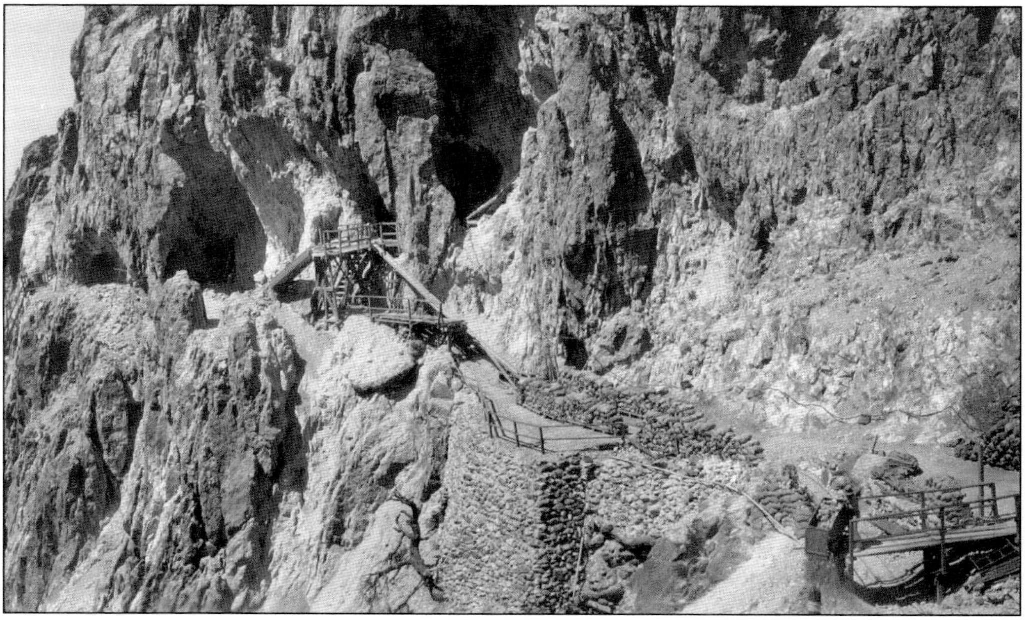

Kearsarge, one of the earliest towns in the region, was built above Independence, below Onion Valley, as a base for mines at nearly 12,000 feet on Kearsarge Peak. The district was built and named during the Civil War by Union sympathizers. Supposedly, 1,500 people lived here in 1865. In the winter of 1866–1867, an avalanche swept through the town, and many died. This photograph was taken in 1871. (Courtesy of ECM.)

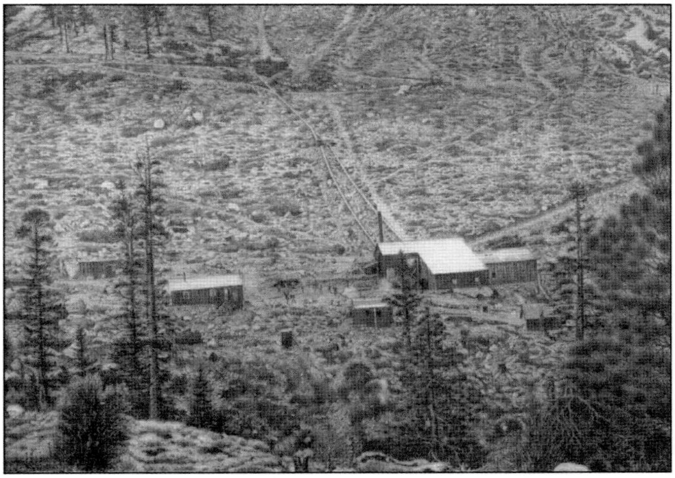

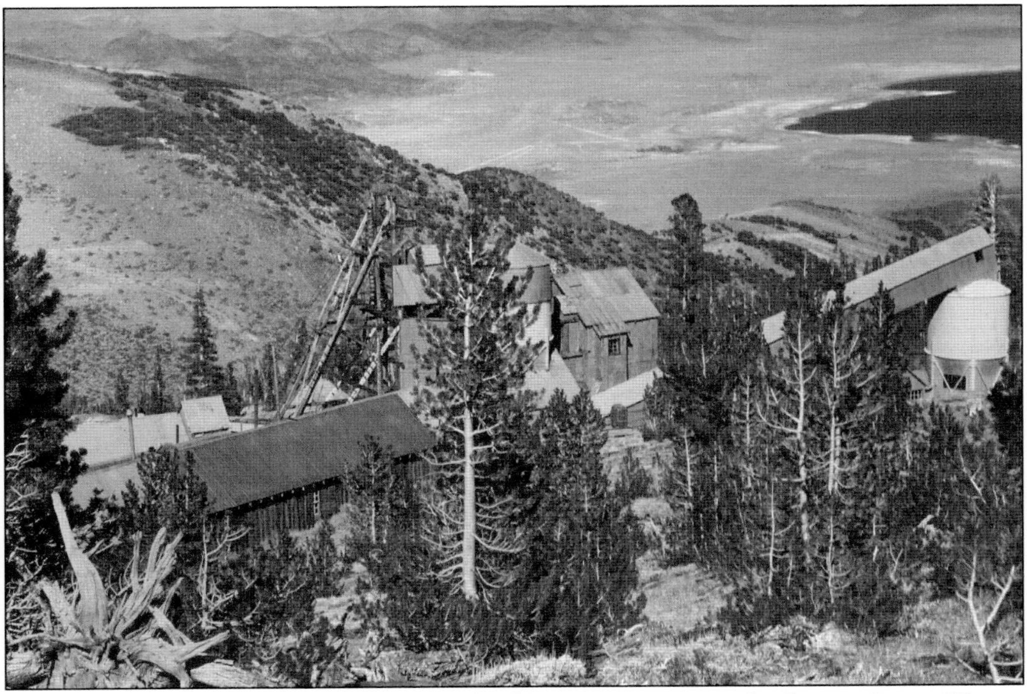

The Log Cabin Mine was the last substantial gold mine to operate in the Eastern Sierra. Way up at 10,000 feet on Warren Bench above Mono Lake, the operation was never easy to get to. In 1930, the Mutual Gold Company bought the original "diggins" and then built most of the structures. The operation included a mercury-processing mill, and when the gold bricks were brought down the hill, the Lee Vining Post Office shipped them straight to the federal repository at Fort Knox. After World War II, the operation continued for a few years, but as the ore diminished and the price of gold stayed fixed at $35 an ounce, it quit producing in 1948. (Courtesy of Frank Cassidy.)

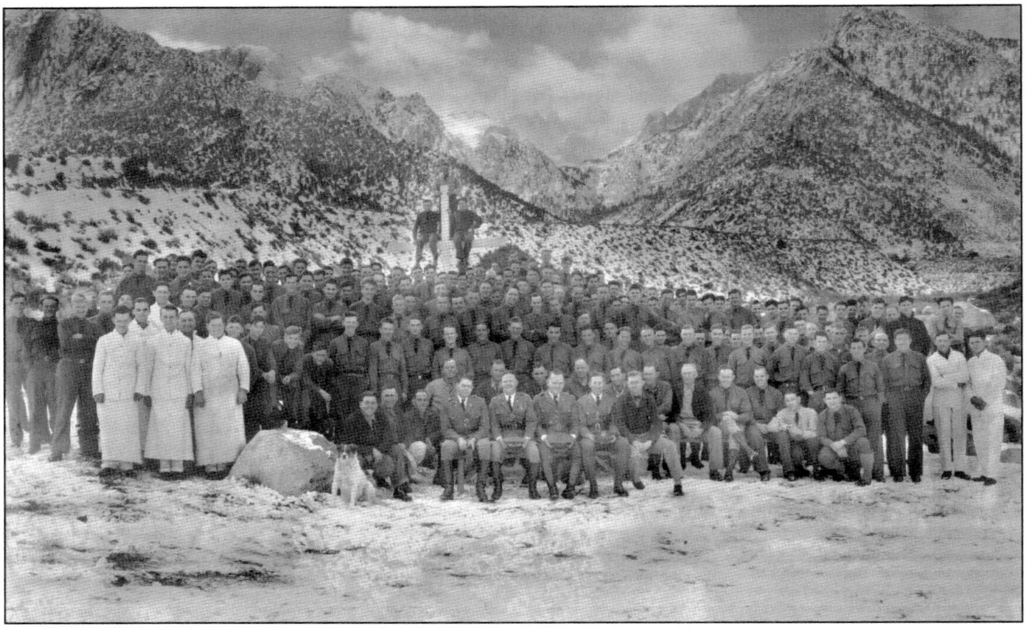

During the Depression, the Roosevelt Administration put 9.5 million young men to work at hundreds of projects with the Civilian Conservation Corps. Teams of "Cs" came to the Inyo in June 1933. State welfare agencies collected workers, the Forest Service designed projects, and the Army ran camp logistics. In California, Brig. Gen. "Hap" Arnold was in charge. The main camp on the Inyo was set up near Lone Pine, and about 200 workers lived at what is now Lone Pine Campground. Here, the group poses on Easter below its most difficult and proudest project, the building of the road from Lone Pine to 8,300-foot-high Whitney Portal. Below, a CCC bulldozer drags a "Fresno scraper" to smooth the Whitney road. The worker riding behind steered the trailing weight. (Allan Ramsey photographs, courtesy of ECM.)

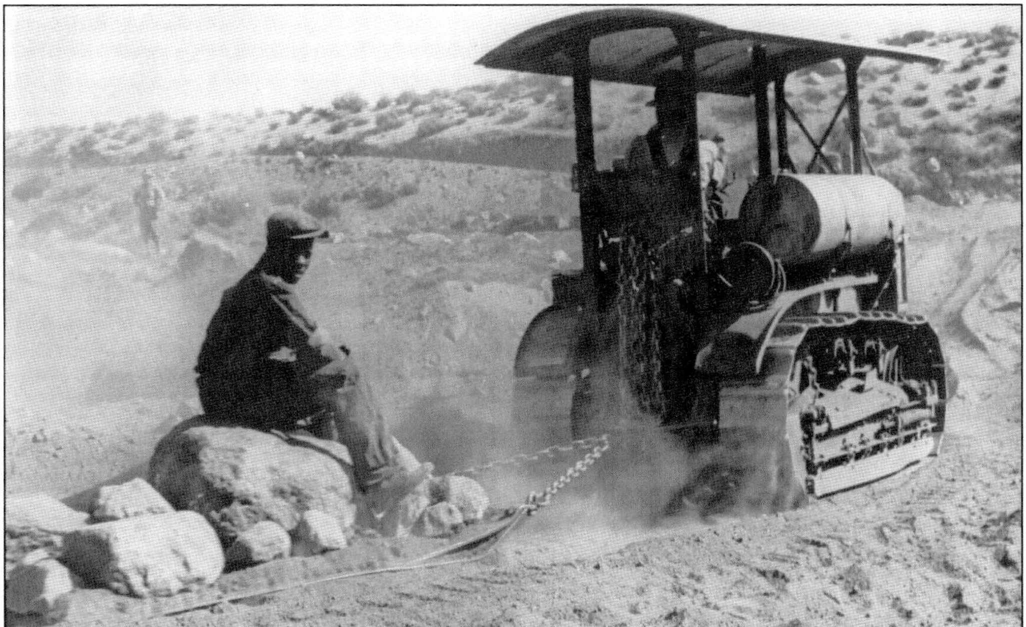

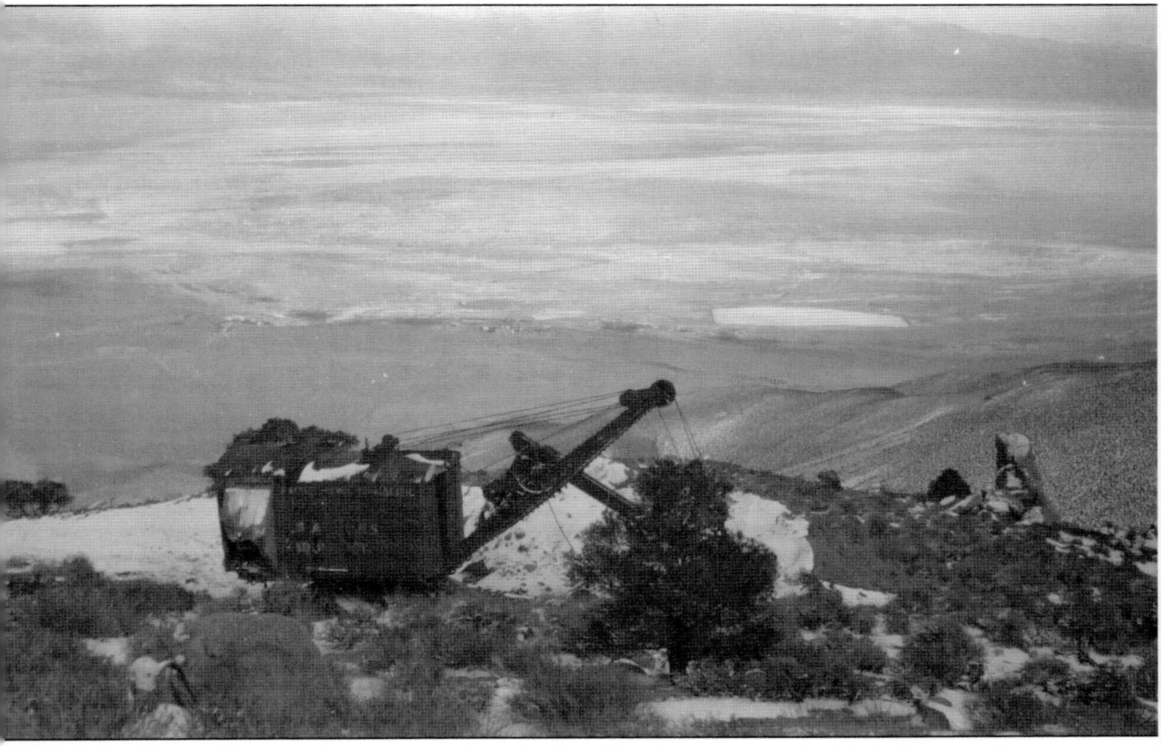

Road engineers and commercial interests hoped to build highways across the High Sierra in three locations south of Tioga Pass. The southernmost plan was for a road from south of Lone Pine to Porterville. Construction for this started around 1930, and a "gas-electric shovel" is shown here excavating switchbacks up the impressive grade out of the Owens Valley. The Forest Service went along with the plan, but the project bogged down in expenses. In later years, the Sierra Club and others were successful in putting this plan to rest with wilderness designation. Around 1970, these switchbacks were paved, and now, this road offers smooth access to the 10,000-foot-high trailhead into the Cottonwood Lakes basin. The other two proposals would have linked Kings Canyon to Bishop and Fresno to Mammoth. (Courtesy of ECM.)

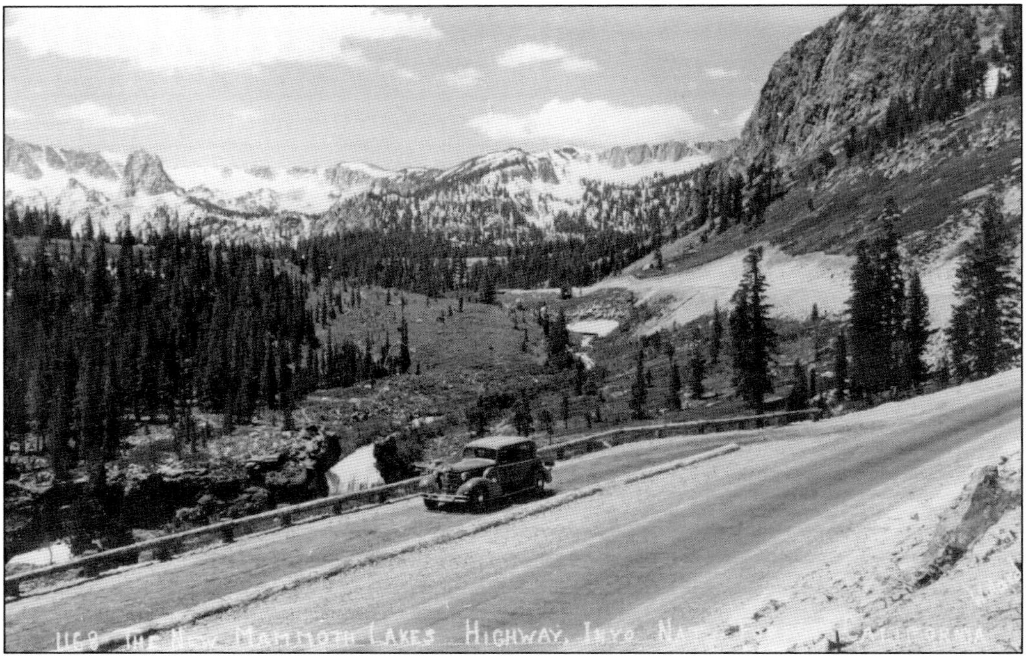

This high-standard road was built in 1939 into the Mammoth Lakes basin. Below the site of this photograph, the road veered away from "Old Mammoth," initiating a new town district. Above here, the road climbed to Lake Mary and continued to Horseshoe Lake. It was later labeled Highway 203 and was slated to extend over Mammoth Pass to Fresno. In the 1960s, the development of Mammoth Mountain Ski Area prompted the construction of a changed Highway 203 to serve the ski area and continue over Minaret Summit toward Fresno. (Courtesy of ECM.)

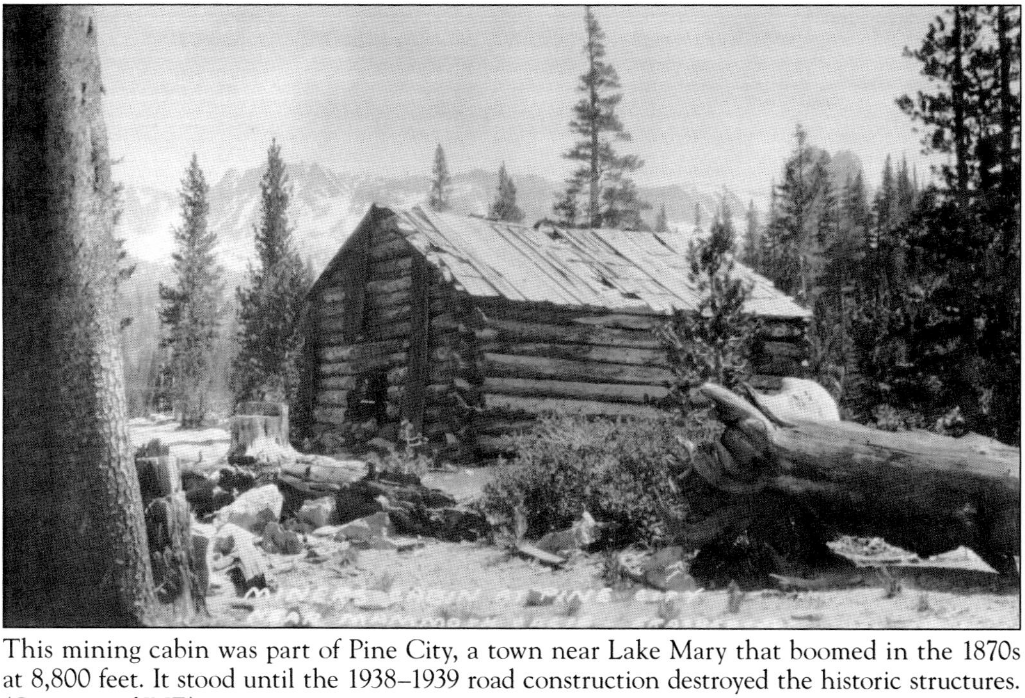

This mining cabin was part of Pine City, a town near Lake Mary that boomed in the 1870s at 8,800 feet. It stood until the 1938–1939 road construction destroyed the historic structures. (Courtesy of INF.)

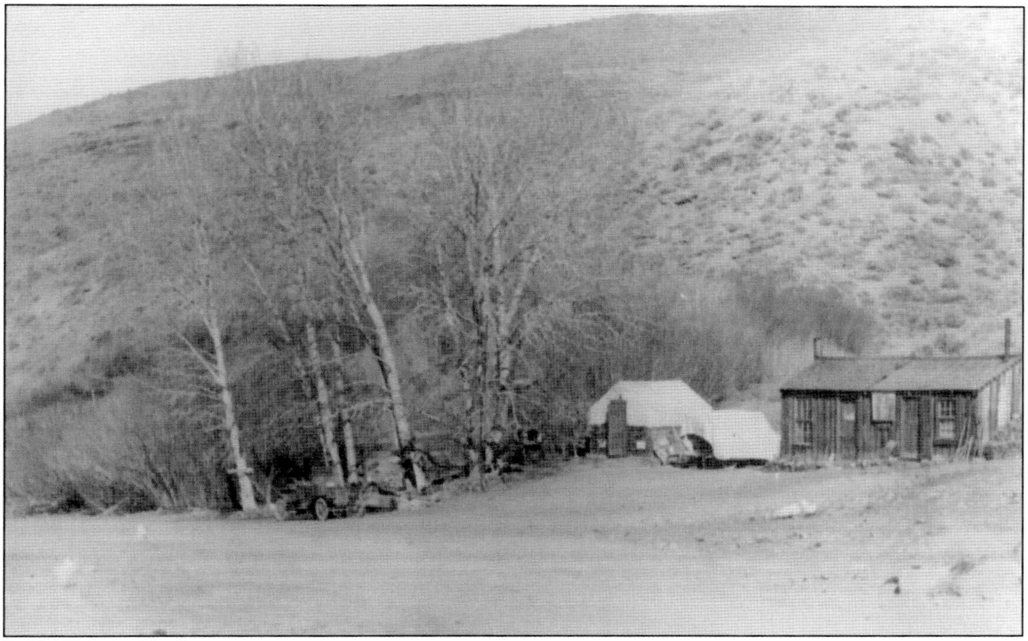

For decades, the site in the photograph above was the tollbooth on the "Deep Springs Valley Toll Road." Built by three entrepreneurs in the 1860s, this was the only reasonably motorable track across the White and Inyo Mountains for a century. In 1913, the American Automobile Industry launched an expedition from Indiana, hoping to discover and promote the best route to the Golden State. The leader A.L. Westgard brought travelers to this road, and the 7,440-foot-high pass across the Whites now bears his name. This track operated as the last privately owned toll road in California until 1921. Below, an early roadster that is well equipped for backcountry travel pauses at the sign at Westgard Pass. (Above, courtesy of INF; below, courtesy of ECM.)

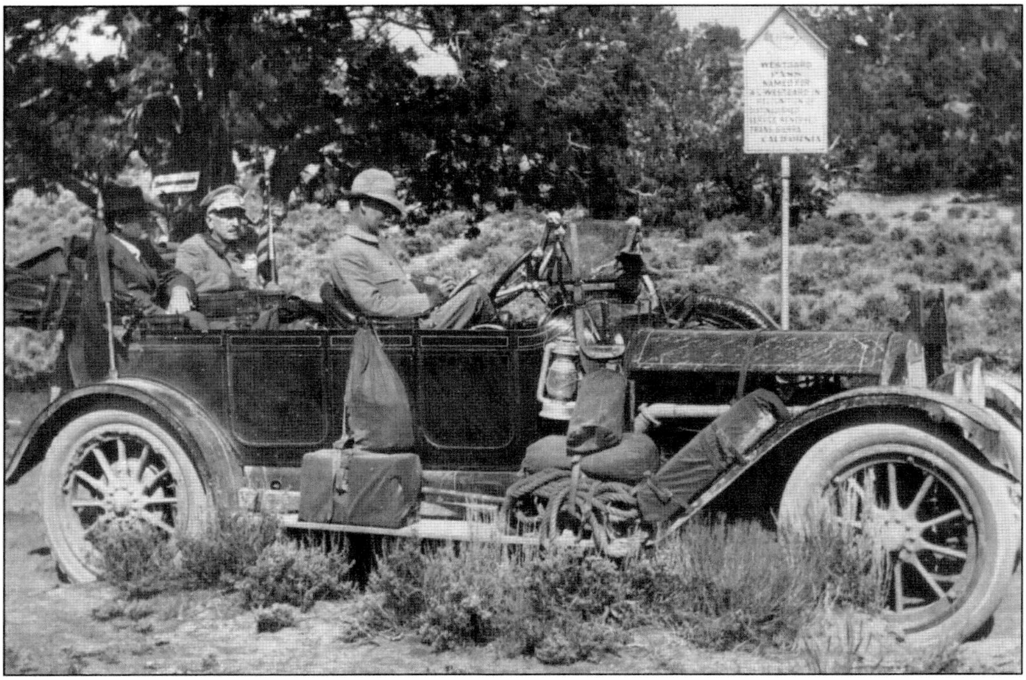

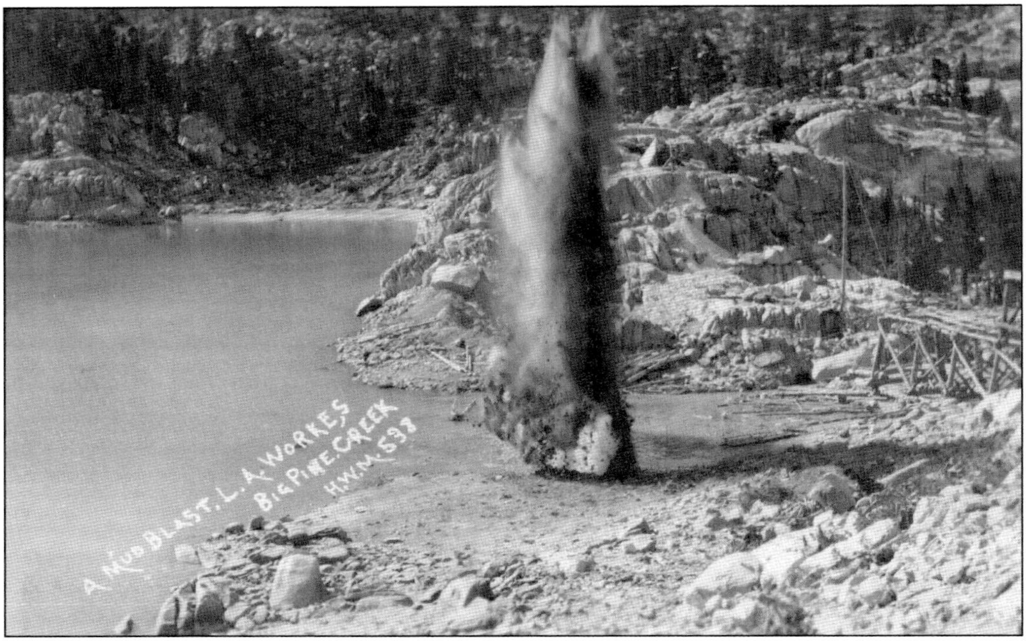

In the 1920s, the Los Angeles Department of Water and Power (DWP) wanted to augment its operations with hydroelectric power, and it chose Big Pine Canyon. Since 1895, Big Pine farmers had enhanced autumn flows into the canyon with a gated tunnel they had blasted beneath the waters of Second Lake, found at 10,000 feet beneath the Palisades. DWP decided to enlarge that capacity with a larger tunnel and a dam. The photograph above shows one of the blasts to enhance the tunnel. To gather rock to fill the dam, they constructed a short rail system with two Fordson locomotive tractors and a large air compressor to power a rock drill. Tons of iron equipment were hauled up to the site on mules, often hitched in pairs with pivoting saddles. However, the Big Pine Lakes were already well loved as some of the most scenic in the Sierra; the operation attracted "negative attention" and was never completed. DWP shifted its hydroelectric focus to the Owens Gorge and the creation of Lake Crowley. Debris remained at Second Lake until it was ferried out by helicopter in 1976. (Courtesy of ECM.)

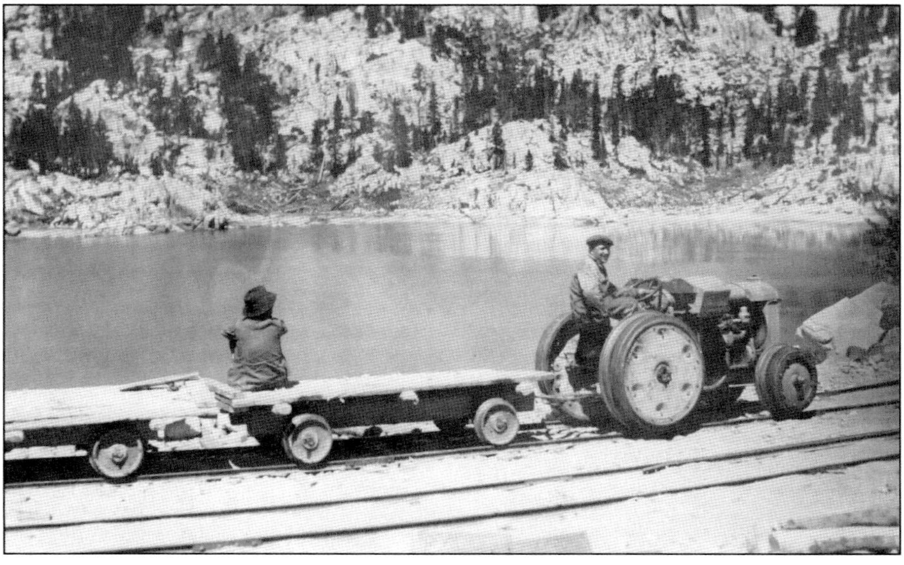

Three

Bounty for Hardy Individuals

Hunting and Fishing

Throughout much of the 20th century, the term "outdoor recreation" was known to mean gathering the fruits of a wildland chase, either by hook and line or by bullet. Hunters have long sought the Inyo as one of California's most exciting regions for deer. But the activity that has spawned the most High Sierra dreams—and family visits—is trout fishing.

Most of the Sierra's steep, recently glaciated drainages had no native fish. However, the meadowy headwaters of the Kern River's south fork and Golden Trout Creek spawned one of America's most legendary fish species, the golden trout. This trout evolved on the upper Kern beginning about 170,000 years ago, when volcanic flows created a waterfall barrier that isolated steelhead rainbow trout that had migrated up and down the river for millennia. The highland rainbows evolved into a smaller, stunning subspecies with golden and red bands, likely because the coloration makes it harder for predators to see them in the shallow streams with rusty-yellow gravel.

Pioneering visitors and managers made it a priority to bring trout to the rest of the Inyo's high country, usually in milk cans strapped onto mules. As early as 1876, fishermen carried golden trout north into the Cottonwood Creek drainage. By 1917, the California Department of Fish and Game established a spawning station at Cottonwood Lakes to gather golden trout eggs for rearing at the new Mount Whitney fish hatchery near Independence. But rainbow trout were proving to be the easiest, fight'n'st fish for dispersal, and the hatchery also started rearing them. They began supplying rainbows and goldens to just about anyone who would take them up to the high country.

In the early 1920s, Forest Supervisor Tom Jones was a fishing fanatic. He had a mission to create the rainbow trout capital of the West. Jones organized the Rainbow Club, an association he described as having "over 3,000 members, including many nationally important people. The club was planting 2.5 million fry annually . . . [in cooperation with] both Inyo and Mono counties, who appropriated $2,000 per year."

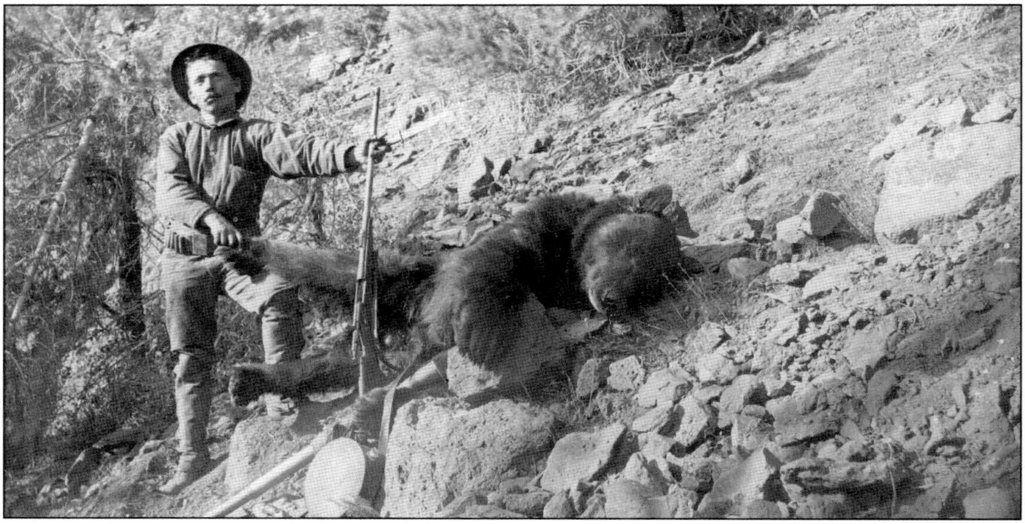

Sheepman Pete Giraud poses with the grizzly he shot in 1897 at Templeton Meadows on the Kern Plateau. It is thought that prehistoric California had one of North America's densest populations of extra-large grizzlies. The bear—predator of livestock and a danger to people— was not tolerated, and the last living icon on the state flag was shot in 1922. The elimination of the grizzly was fundamental to the modern concept of the Sierra as a gentle mountain wilderness. (Courtesy of ECM.)

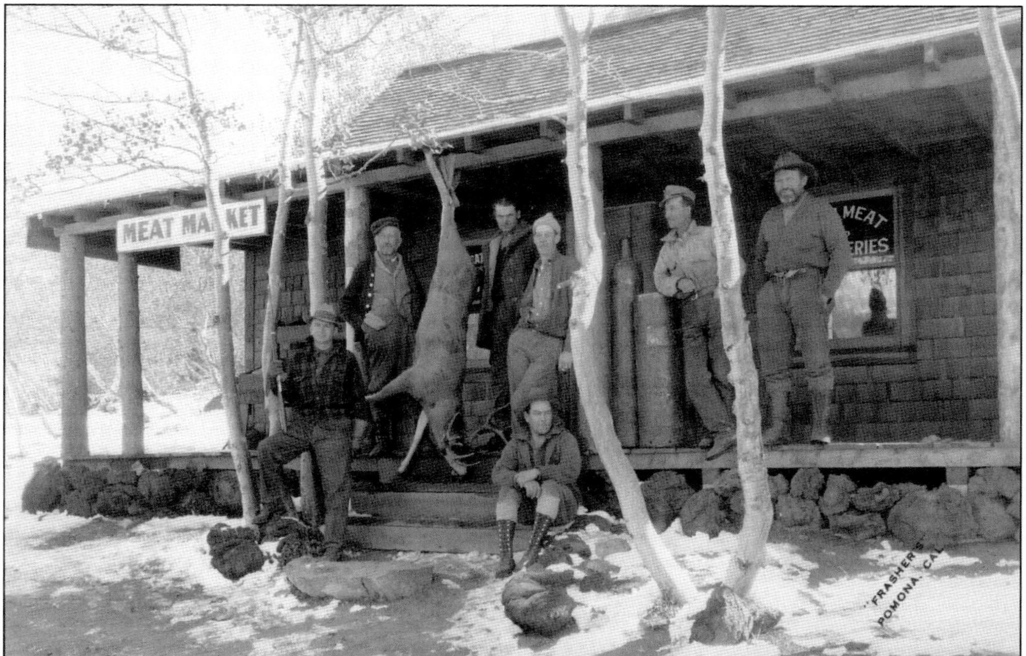

Deer hunters in the 1920s gather with a five-point buck at a store in June Lake. The Sierra mule deer, a subspecies of white-tailed deer, are named for their large ears. Most of them migrate to high country meadows after the passes clear and return to winter ranges in the fall. Until recently, the fall hunting season was one of the busiest times for recreation on the Inyo. The Inyo and California Department of Fish and Game still pay close attention to deer populations, hunting, and potential herd impacts. (Courtesy of ECM.)

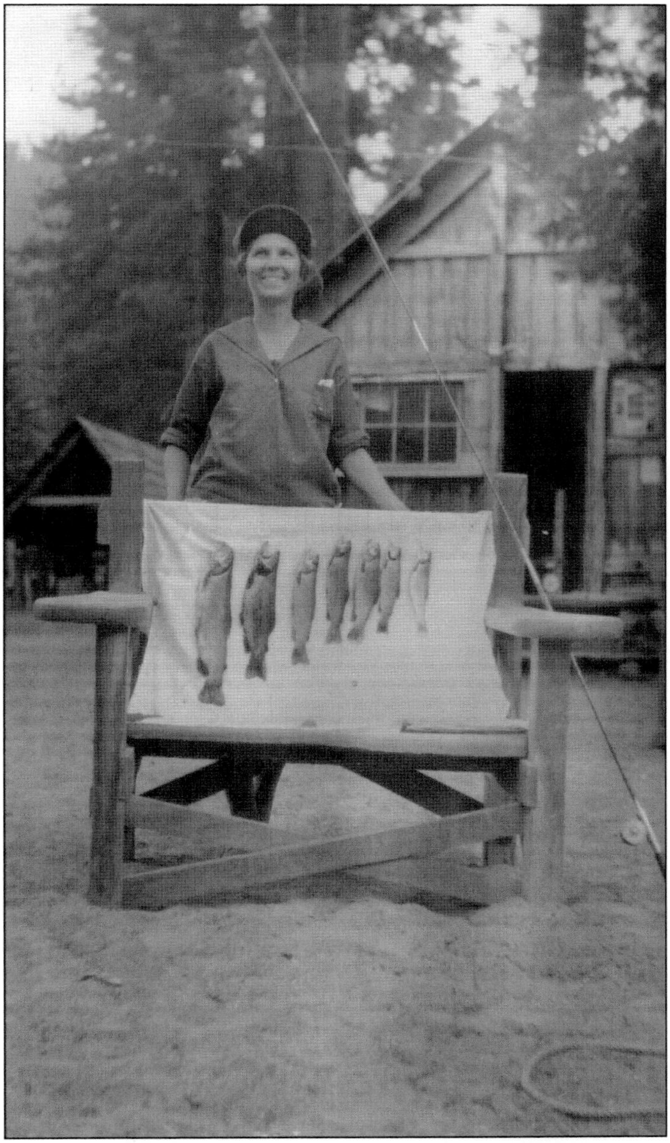

A woman of the early 1900s beams with trout caught on her fly rod near Jordan Hot Springs. Jordan was a resort near Ninemile Creek, a tributary of the Kern River. Before car vacations came to dominate America, it was a well-known backcountry resort. It took a couple of days to ride in, but customers raved about soaking in hot springs after trout fishing within easy walking distance of the camp. (Courtesy of INF.)

A train of 20 mules commissioned by the California Department of Fish and Game carries live golden trout over 11,400-foot-high Piute Pass for planting into the lakes of Humphreys Basin. This picture was taken on July 26, 1914, and the snow was still six feet deep at the pass. (Courtesy of ECM.)

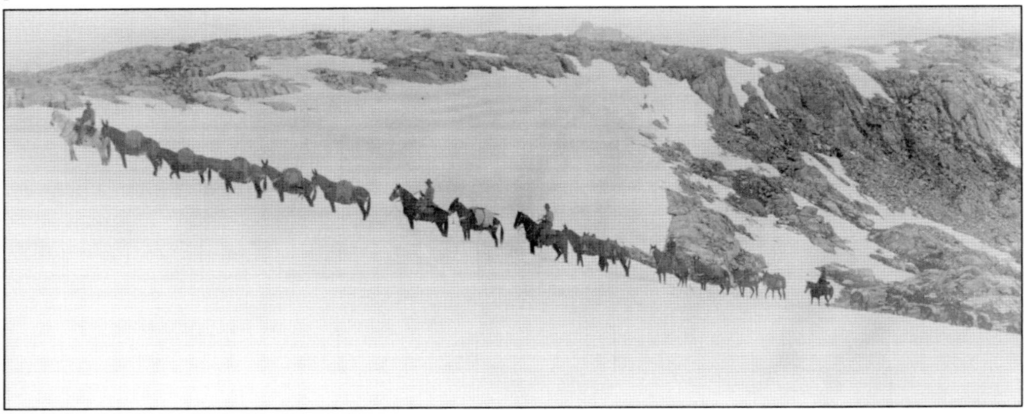

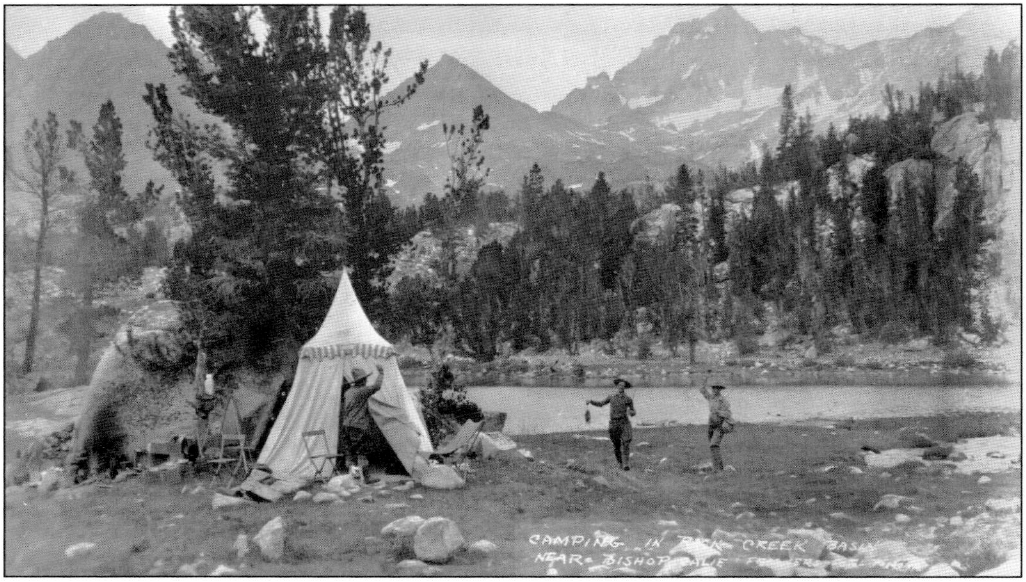

A pair of fishermen returns with their catch to a high country campsite at Heart Lake in Little Lakes Valley, the upper basin of Rock Creek. This basin has many 10,000-foot-high lakes planted with rainbow trout. A mining road for relatively easy access and classic panoramas of 13,000-foot-high peaks made this one of the most popular destination areas on the Inyo. After the 1964 Wilderness Act, the old, very rough mining road was closed to vehicles and made into a trail, but the basin has remained as popular as ever. Below, gathering around a campfire after a day's wilderness adventures is a timeless experience that many people have enjoyed in the Sierra. (Courtesy of ECM.)

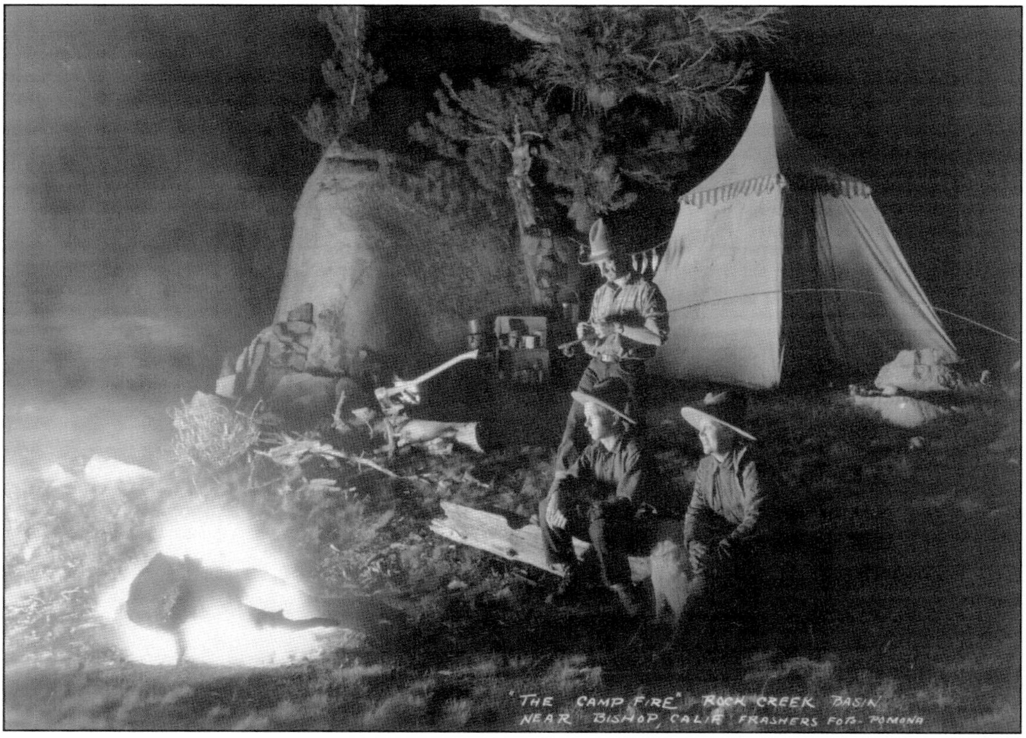

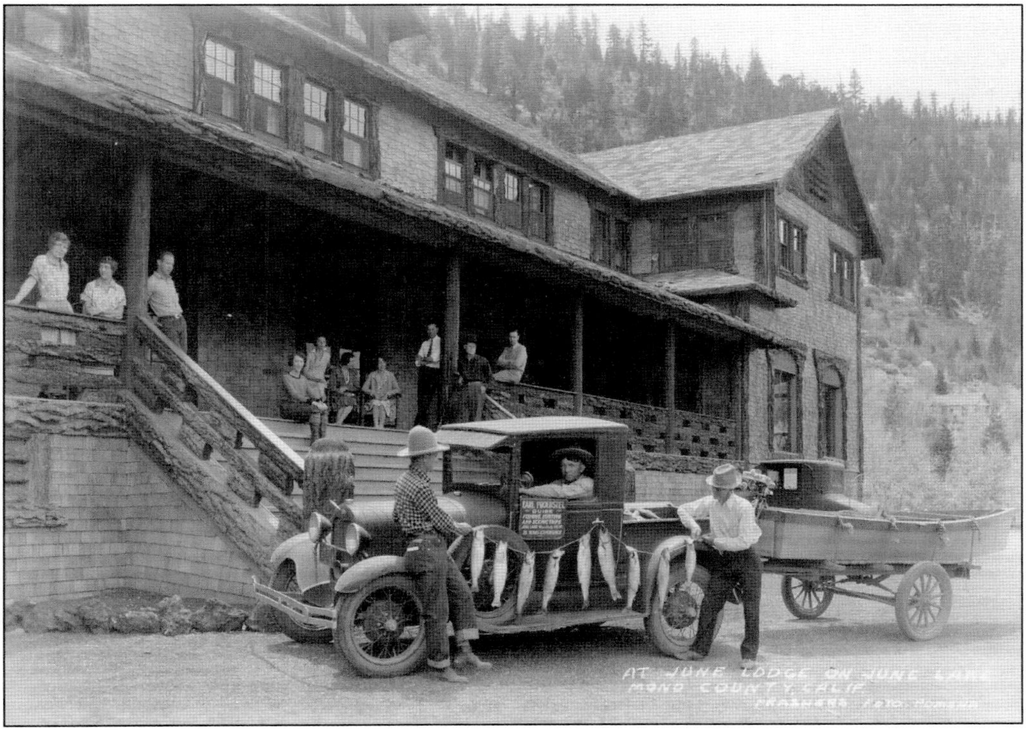

Lakes along the easily accessed June Lake Loop quickly became some of the premier fishing destinations in California. Photographer Burton Frasher (on fender at left) poses in his own picture with fishing and hunting guide Earl Proebstel (driving) at the stylish June Lodge in the late 1920s. The arc of Reverse and Rush Creeks includes June, Gull, and Silver Lakes, plus Grant Lake reservoir. Carson Camp, a pioneering facility at Silver Lake, was probably the first lodging resort on the Inyo, founded in 1921. (Courtesy of ECM.)

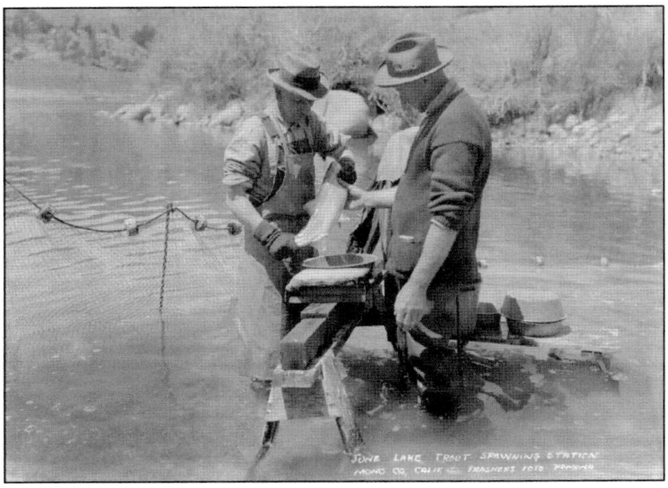

California Department of Fish and Game developed fish-rearing techniques in the June area. Here, two men extract milt and roe from trout in a weir-enclosed area of Silver Lake. Later, in 1926, the division set up a better weir system along Rush Creek and then an early hatchery along Fern Creek, a tributary above Silver Lake. Cutthroat and brown trout from Fern Creek were planted up and down the Sierra for many years. (Courtesy of ECM.)

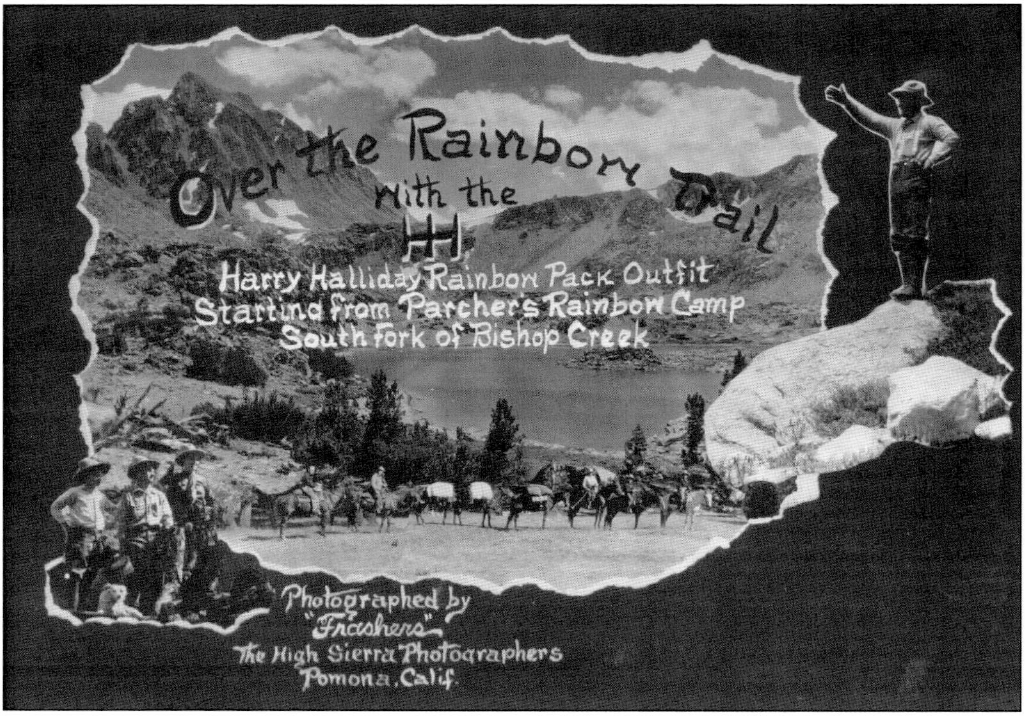

Emphasizing the allure of fishing for rainbow trout in the High Sierra above Bishop, this 1920s Frasher postcard advertises Harry Halliday's Rainbow Pack Outfit and Parcher's Rainbow Camp. From Rainbow Camp, one could ride up "the Rainbow Trail" over the 11,200-foot-high Bishop Pass and descend to Rainbow Lakes. The scenery includes 13,893-foot-high Mount Agassiz, left, reigning over the pass. Below, the Rainbow Club organized this display of trout taken from the Mono (now Inyo) National Forest. The club was an organization of civic leaders and fishing enthusiasts who promoted and organized fish stocking on the Inyo. (Courtesy of ECM.)

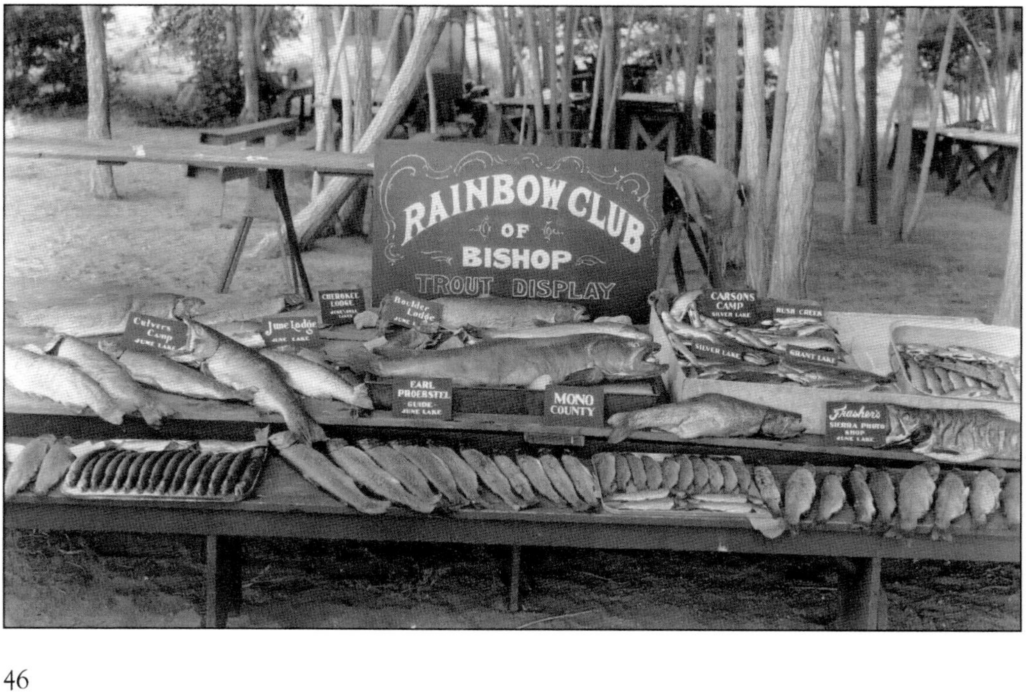

Gardisky's Camp Tioga, near Tioga Lake, was the highest early resort on the Inyo, at 9,600 feet. Today, it is called Tioga Pass Resort. Each spring opening has brought an adventure for the owner in wondering how the place fared through the winter. Recently, the place became well known among backcountry skiers for staying open through the winter by snowmobile access. But challenges in complying with regulations for fuel storage and water supply have again pressed the owner to close each winter. At right, Burton Frasher nets a trout out of nearby Tioga Lake. (Courtesy of ECM.)

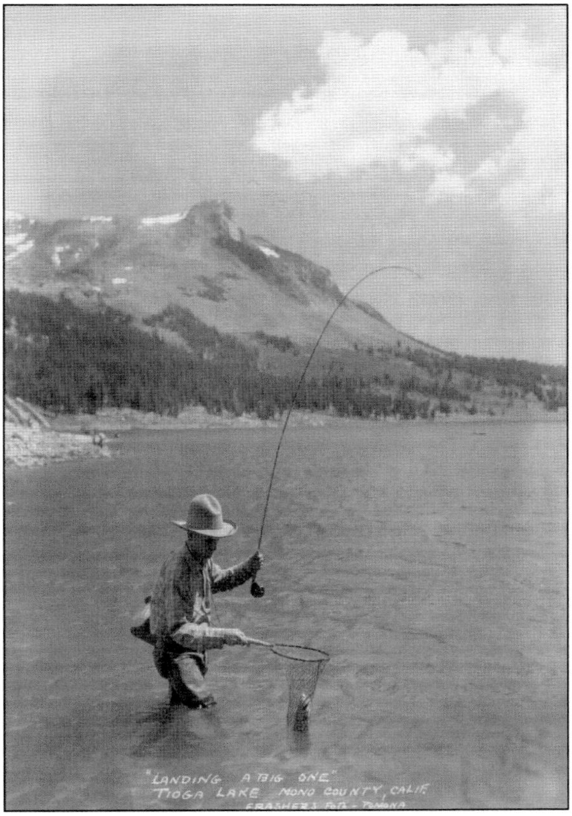

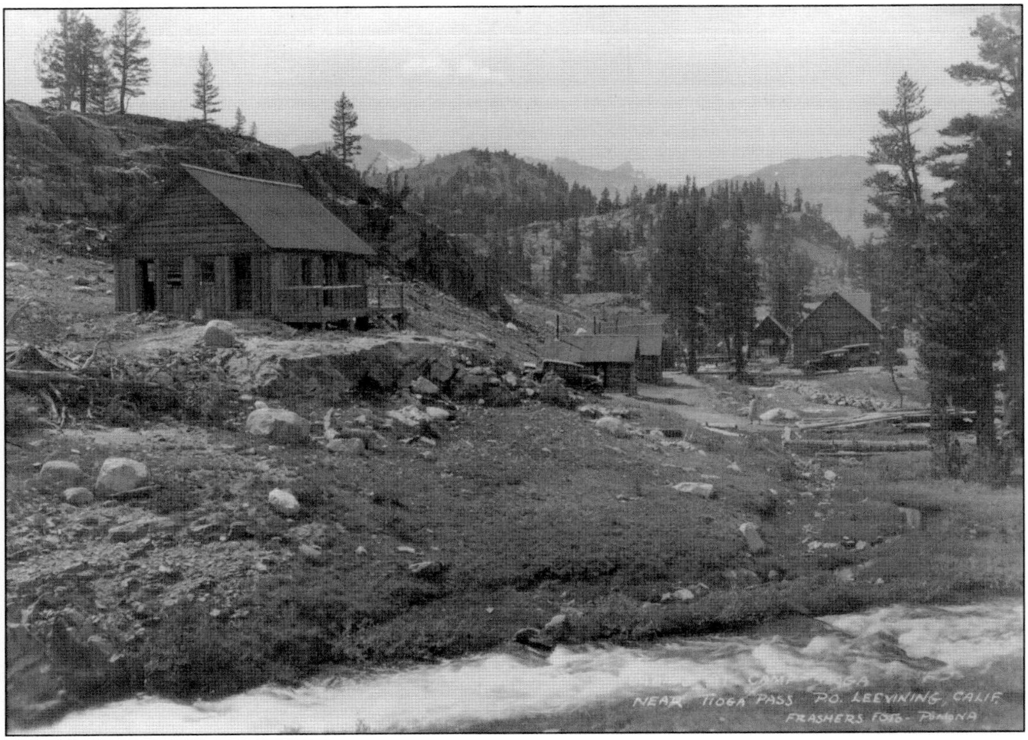

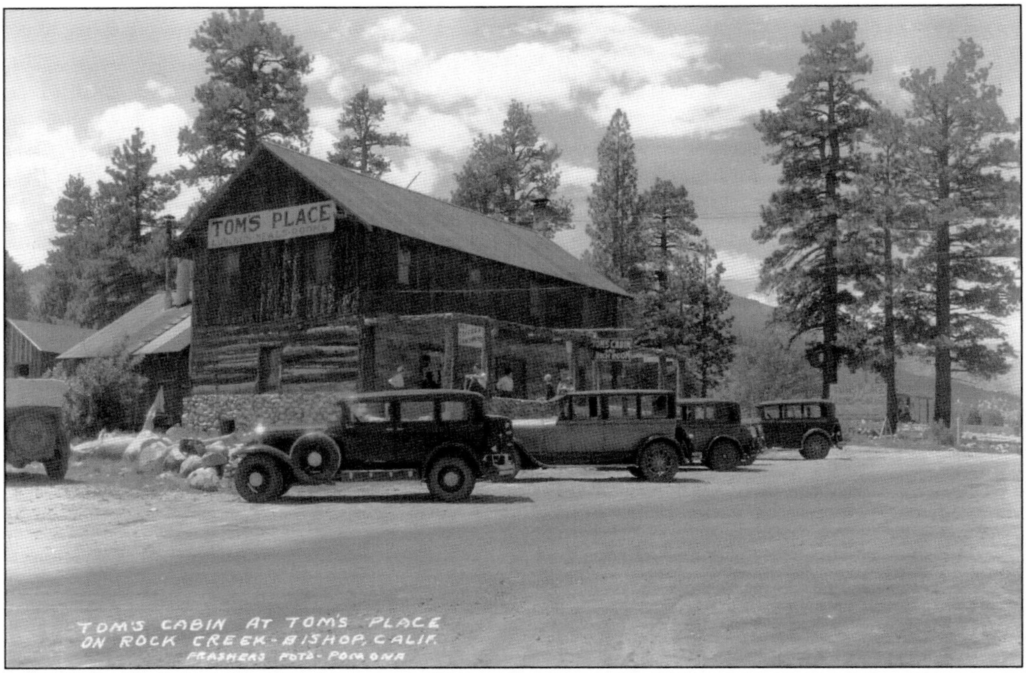

Hans Löf built this lodge at a bench in the long Rock Creek Canyon in 1917. A few years later, he sold it to Tom and Hazel Yerby, who called it Toms Place. They added a garage and gas pumps that were welcome to drivers arriving at the climb to 7,000 feet atop Sherwin Grade. This building burned down in 1947, and the current lodge was built. Throughout its history, Toms Place has especially thrived on the opening day of fishing season in late April. A restaurant and bar were added across the road, and the compound is still a popular landmark under forest service lease. (Courtesy of ECM.)

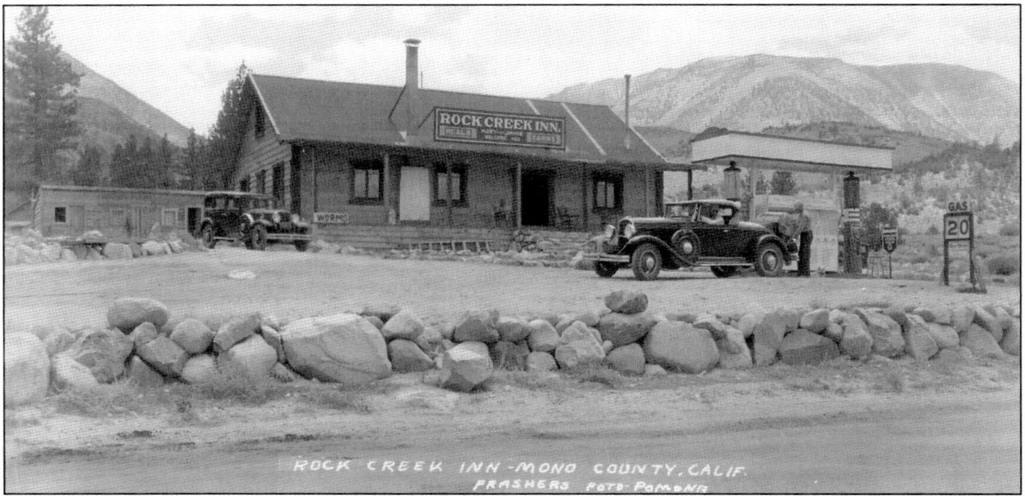

In the early decades of recreation on the Inyo, there were many small resorts like this along Highway 395. Jim and Frankie Birchim built the Rock Creek Inn in 1933, a few hundred yards west of Toms Place. This building burned down in 1945, but a new owner built a more extensive resort, including 26 cabins and an outdoor dance floor. In 1962, it burned down, too. With highway improvements and faster driving, the resort scene since has steadily consolidated in Bishop and Mammoth. (Courtesy of ECM.)

Four

Where Wheels Cannot Go

The Backcountry

Civilization has been built on the apparently logical premise that road access is a boon and, now, a necessity. However, society seems to seek antidotes to all that is built, and important parts of us thrive in places where songs of nature both soft and hard fill the air. The steep ramparts of the Inyo of course resist the call of the wheel, and even as the asphalt carpet was rolled out to most every corner of California, the Sierra, White, and Inyo ranges remained quiet, lofty refuges where feet and hooves are still the way to travel. Since the founding of America, some of its most visionary leaders have articulated how there is something vital to be found in such wild places, where people adapt to the land instead of transforming it to their convenience. The Forest Service began to recognize this in 1929 by designating primitive areas here in the High Sierra and elsewhere. These were areas where few developments other than trails would be allowed.

From early on the call to explore the Sierra backcountry meant a week or a month with a train of pack animals. By the 1930s, a pack station or two to service the public sprouted at every trailhead. Each outfit developed its trails, campsites, and clientele. Veteran "packers" became skilled, hardworking guides to the wilderness and often developed a pride of ownership to their territory.

After World War II, Army surplus and commercial gear of lightweight nylon, aluminum, and down made it more reasonable to shoulder one's own supplies and roam into remote crannies of the range. "Backpacking" the whole John Muir Trail became downright popular. The gear technology evolved as the increasingly mechanized nation saw more need to preserve and visit wild areas, and in 1964, Congress and President Johnson wrote the Wilderness Act and the values of the primitive areas into the national canon. The John Muir and Minarets Wilderness Areas on the Inyo were among the first that Congress designated.

The traditions of backpacking and riding stock into the Sierra evolved out of different generations and cultures, and for some folks they have seemed almost mutually exclusive. While points of contention still arise over trail impacts and traditional rights, the fact is that hooves and boots share the same core values: roaming the High Sierra makes the stress and uncertainty of modern life seem small and the wonders of the natural world seem large.

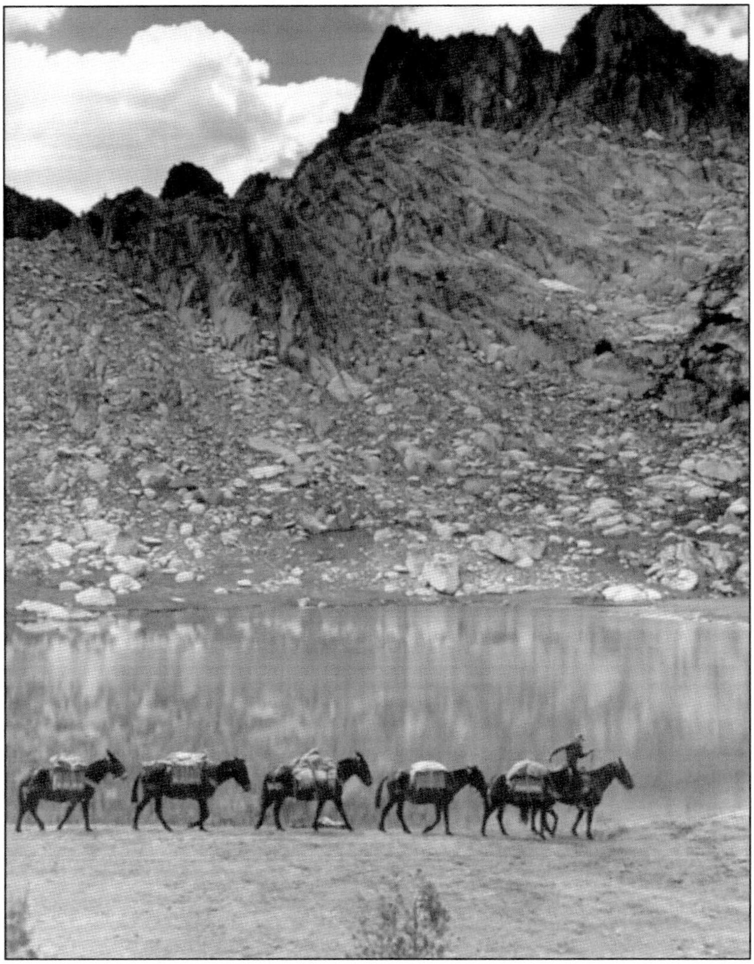

For decades, the way to get into the Sierra backcountry was with mules. Compared to a horse, the sterile cross between a male donkey (jack) and a female horse (mare) will carry more weight (200 pounds or more), persevere over rougher terrain with a steadier gait and stabler personality, and will let a rider know when he is pushing too hard simply by staying put. At left, wrangler Lou Roeser leads a pack train along Squaw Lake on the John Muir Trail. Below, mules were often counted on to carry awkward loads on rough terrain, like this fisherman's dingy. (Left, Steve Lucasik photograph, courtesy of Jim Cooper collection; below, courtesy of ECM.)

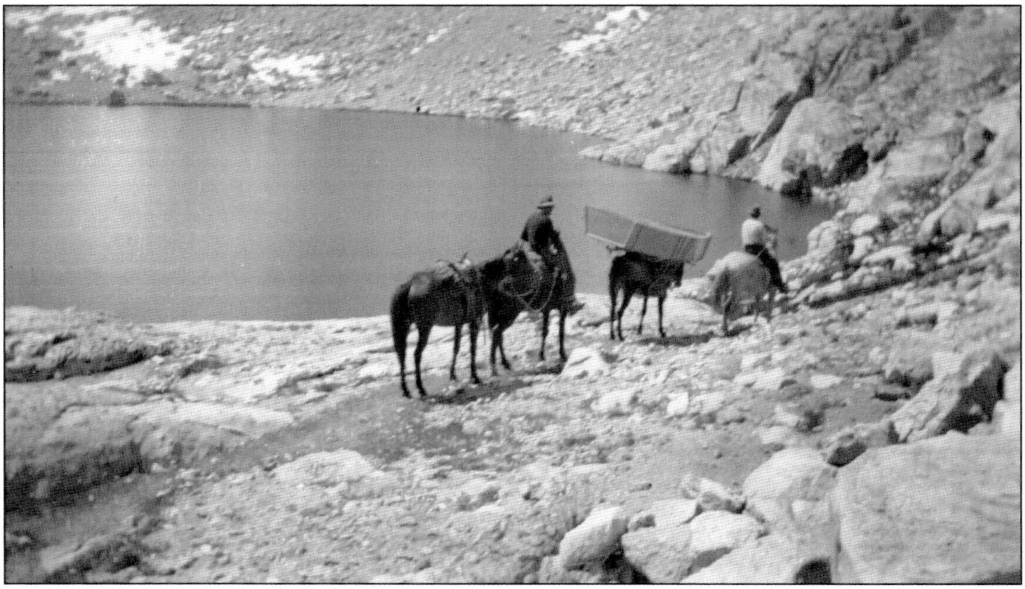

One of the Sierra Club's main goals was to connect people with the Sierra, and for decades, its most successful conduit was the annual High Trip. These became grand expeditions of hundreds of members, who toured for three to five weeks over extensive spans of backcountry. The traveling towns climbed mountains, planted fish, put on plays, and developed a sort of transhumant summer tradition apart from their urban homes. At right is a High Trip camp scene. (Courtesy of Lou and Marye Roeser collection.)

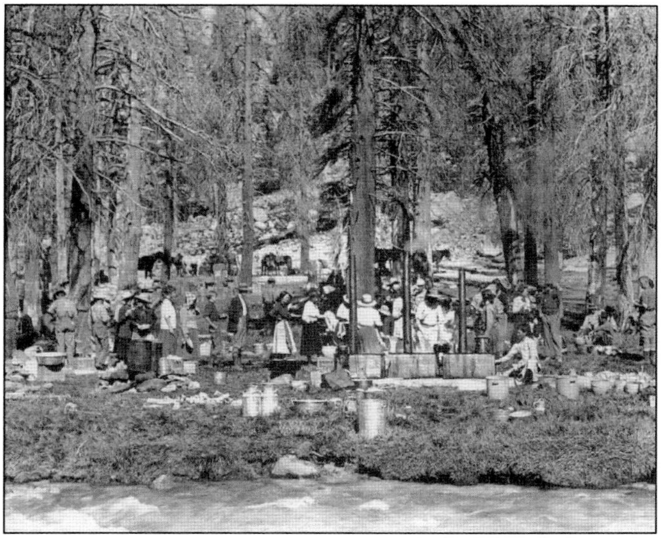

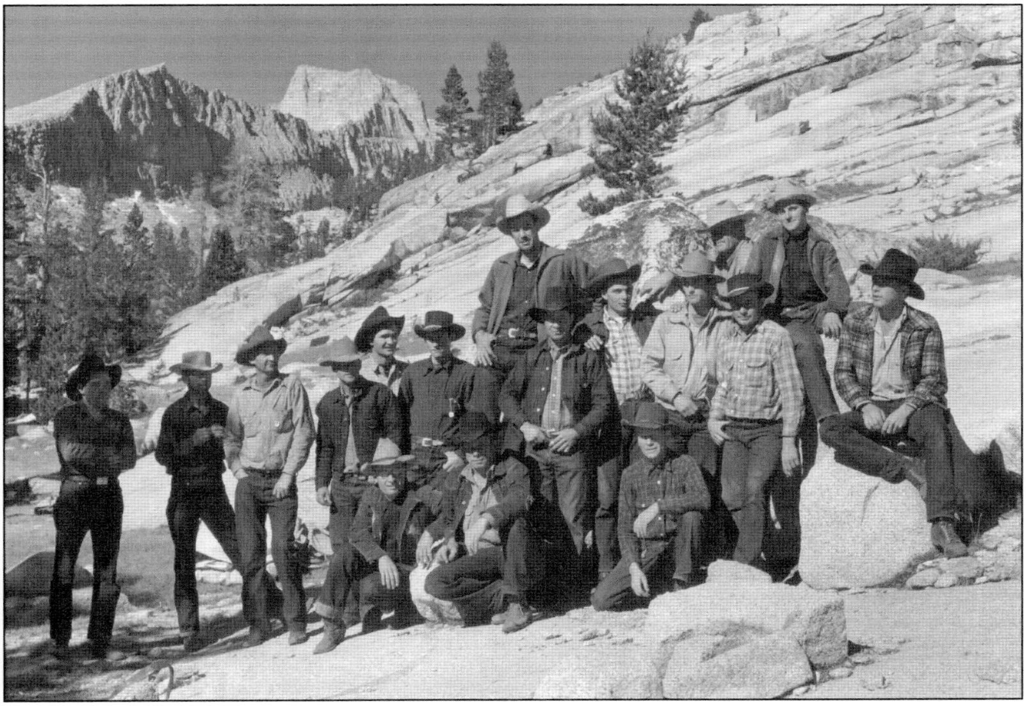

High Trips required many strings of mules, coordinated from several different pack outfits. This is a group of 17 wranglers who served a typical High Trip. The wranglers were mostly ranch hands and others from a rural background. Wranglers and their hiking customers ("footburners") eyed each other with widely varying degrees of respect, curiosity, and sometimes disdain. Norman "Ike" Livermore, standing behind in the white hat, was often lead organizer. As a Stanford graduate and expert horseman, he was a bridge between the two cultures. (Courtesy of Lou and Marye Roeser collection.)

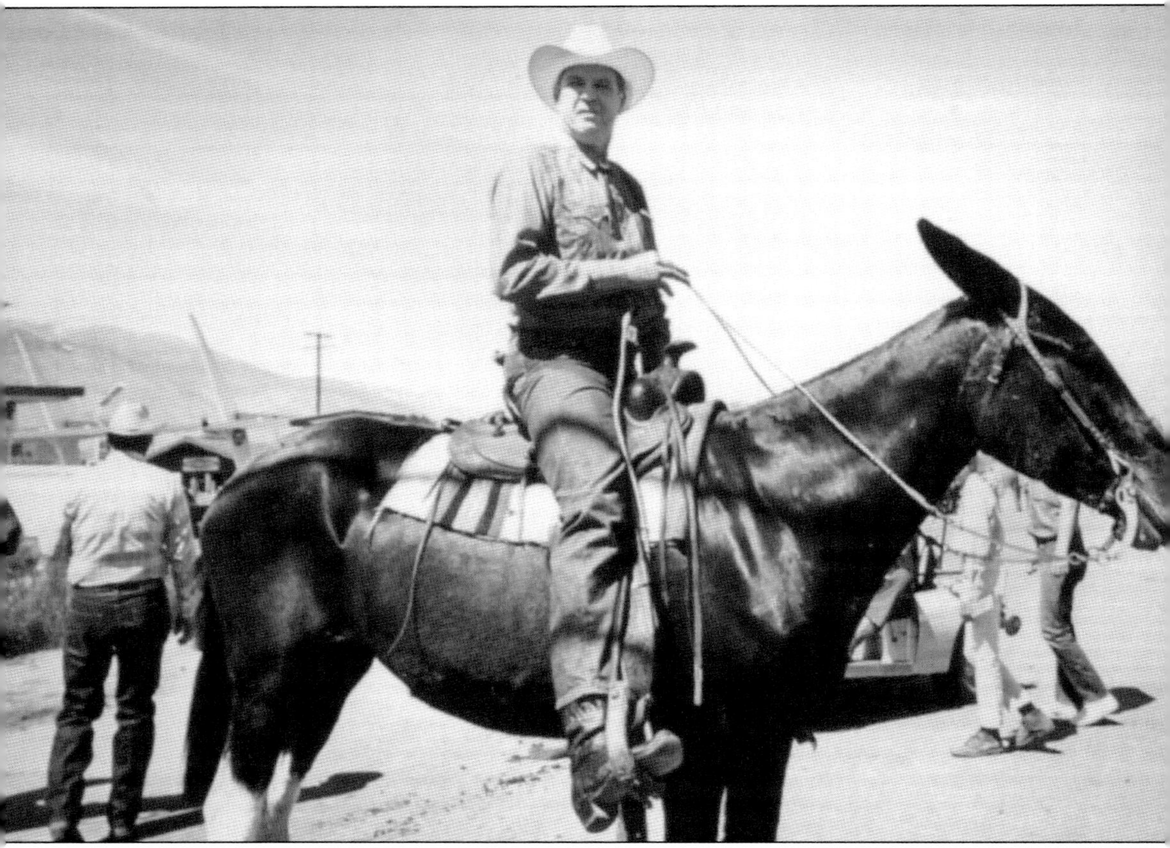

Ike Livermore rides a mule named Elmer as grand marshal in the 1971 Mule Days parade in Bishop. Livermore organized the High Sierra Packers Association in 1935. He was a modern Renaissance cowboy, starting as a teenager, when his activities included riding a horse from his family's ranch in Ojai to Big Sur, and climbing the Grand Teton. He graduated from Stanford in social science and worked summers as a wrangler in Mineral King. In 1936, he finished his MA at Stanford on "the Economic Significance of California's Wilderness Areas," served as executive secretary of the Packers Association, and played catcher on the demonstration baseball team at the Berlin Olympics. He later served on the board of directors of the Sierra Club and as resource secretary for Gov. Ronald Reagan. His friend Martin Litton said, "Ike was a Republican . . . If he had an obsession at all, it was to keep the Sierra Nevada wild for the whole stretch from Tioga Pass . . . south to Walker Pass." (Courtesy of Livermore family collection.)

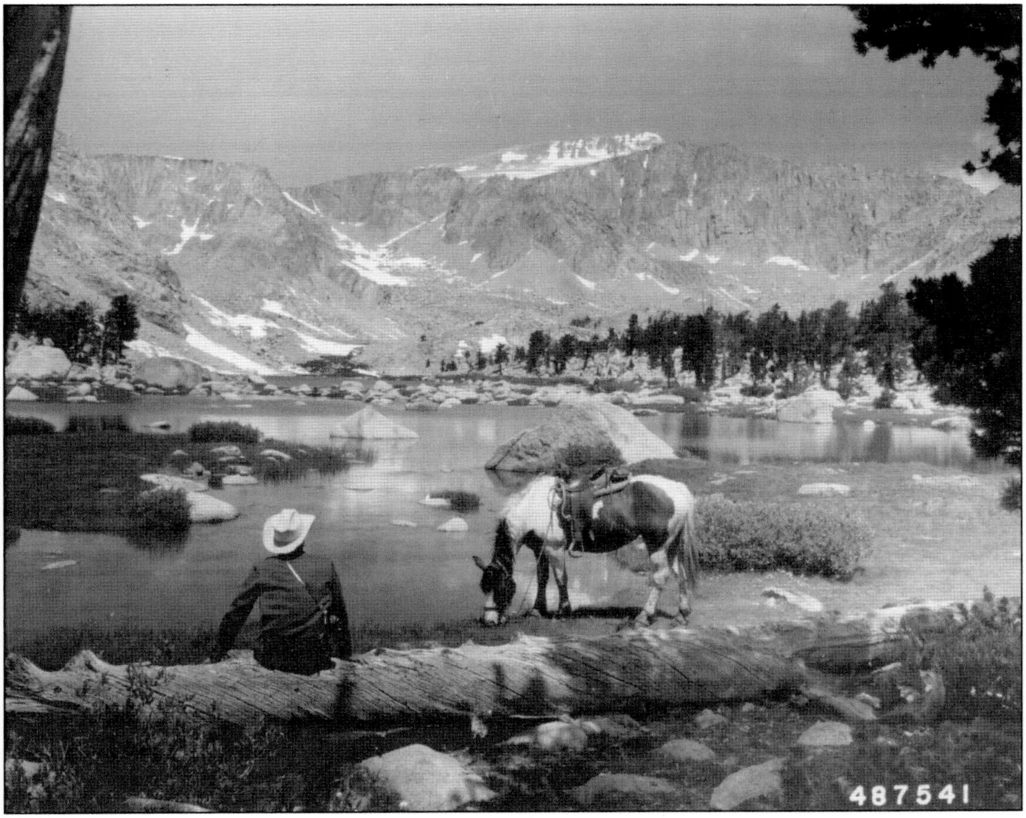

The High Sierra is a "Range of Light" more often than not, but backcountry travelers can also find themselves buried in stressful, even frighteningly harsh remoteness. Above, a rider, possibly an early forest service ranger, pauses with his horse at one of the Cottonwood Lakes below Mount Langley. Today, with the fragile High Sierra a popular land, regulations prohibit grazing in sensitive areas such as this, and in many areas, stock parties are now required to bring in their own feed. Below, a wrangler named Horace Elder gets hit with a snowstorm on top of Mount Whitney while reshoeing his mule for the rocky trail back down. (Above, courtesy of INF; below, courtesy of ECM.)

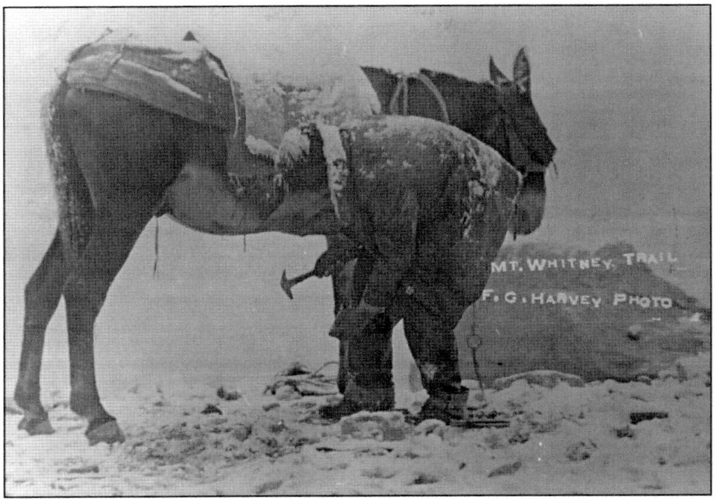

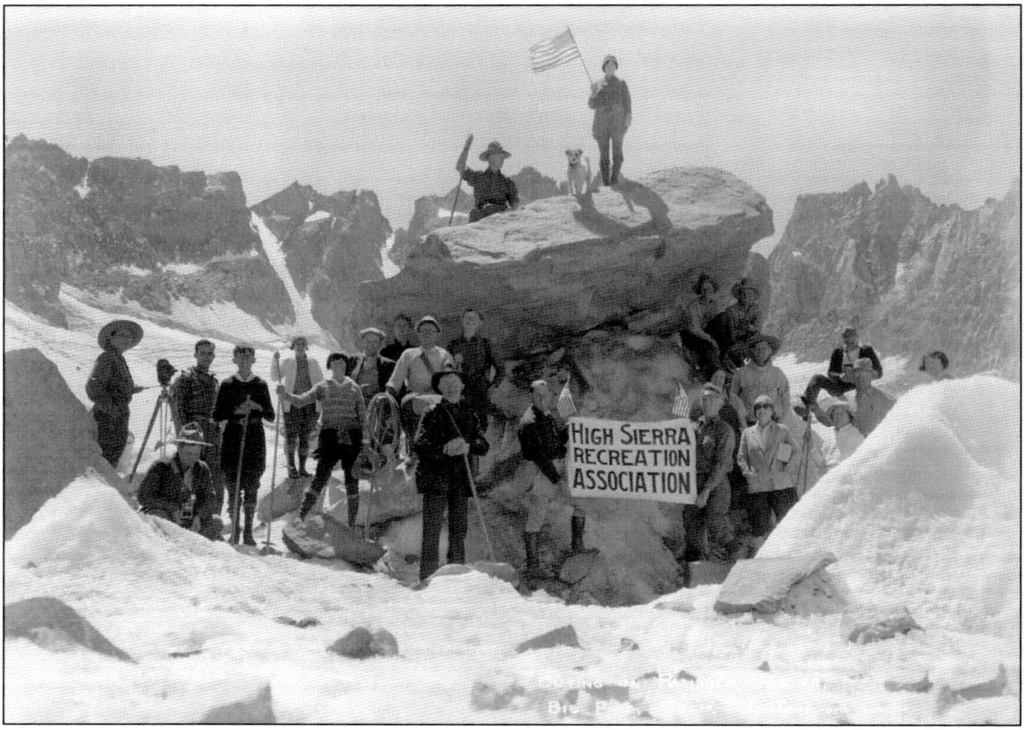

Glaciers were a primary fascination for Alpine travelers of the Victorian era, and the Palisade Glacier—the most distinctive (if still diminutive) glacier in the Sierra—was a great attraction during the first half of the 20th century. The rough path up to its 12,500-foot-high bed from near Third Lake was, and still is, too rugged for stock travel. Here, a short-lived Southern California group called the High Sierra Recreation Association gathers at a common glacier curiosity; a large rock that fell off of a peak and was carried for years on the moving ice. Gradually, it was left perched on the pedestal of snow and ice that it had shaded. (Courtesy of ECM.)

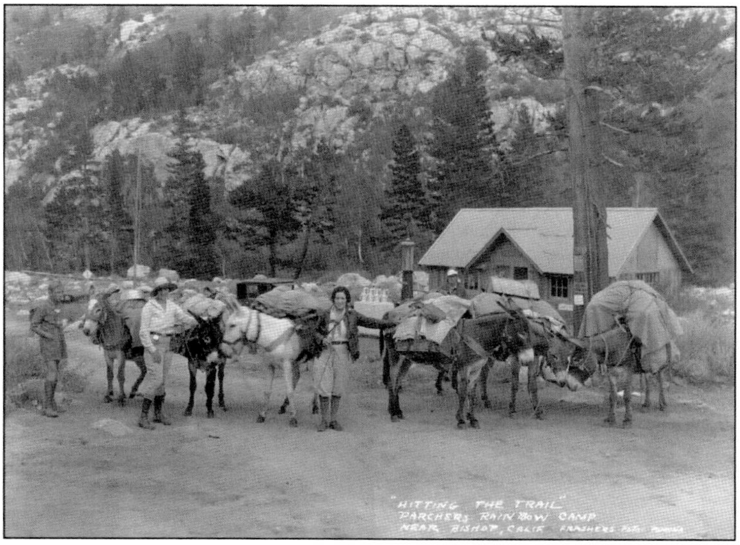

Many Sierra enthusiasts have long preferred to hike, while animals carry their loads. Here, a small group in the 1920s gets ready to walk over Bishop Pass with donkeys at Parcher's Camp, located just below South Lake. The cup ready on the belt for dipping into streams suggests that these are likely Sierra Club members on a privately organized excursion. (Courtesy of ECM.)

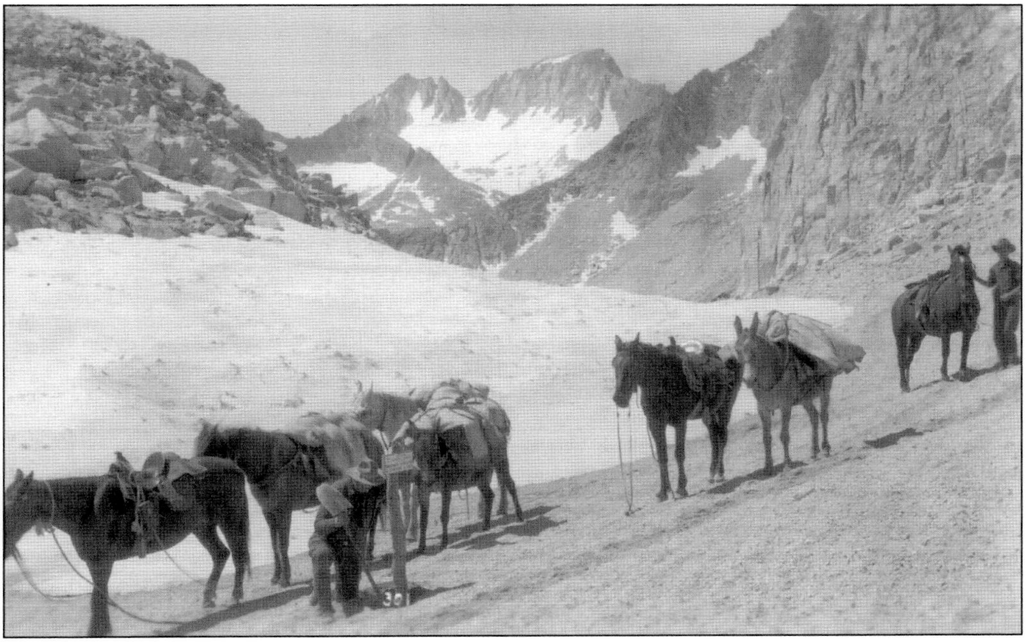

Through the first half of the 20th century, trails were developed over main Sierra passes. Some were created by forest service crews, and many others were established by the pack outfits. Here, a wrangler, probably a forest service ranger, puts in a sign marking the top of 12,000-foot-high Mono Pass, a passage from Rock Creek to the west side at the Mono Creek drainage. This particular route was used by Indians and later by west-side workers building the dam at Lake Edison. They would cross the pass and catch a ride with miners down to Bishop to visit the bars and brothels. (Courtesy of INF.)

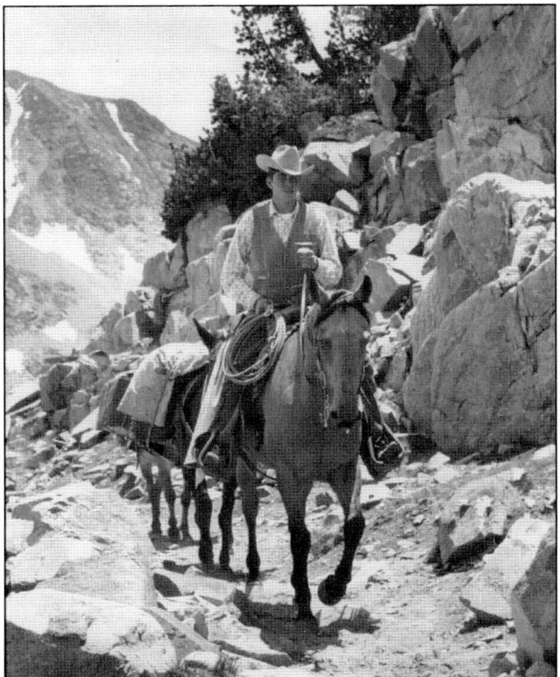

The Inyo employs packers to this day to work on trails, supply crews, and fight fires. Their lead packers are Lee Roeser and Michael Morse, who have developed minimum impact techniques for work in wilderness areas. Here is Lee Roeser at his earlier job, at his parents' Mammoth Lakes Pack Outfit at Duk Pass in 1971. (Courtesy of Lou and Marye Roeser collection.)

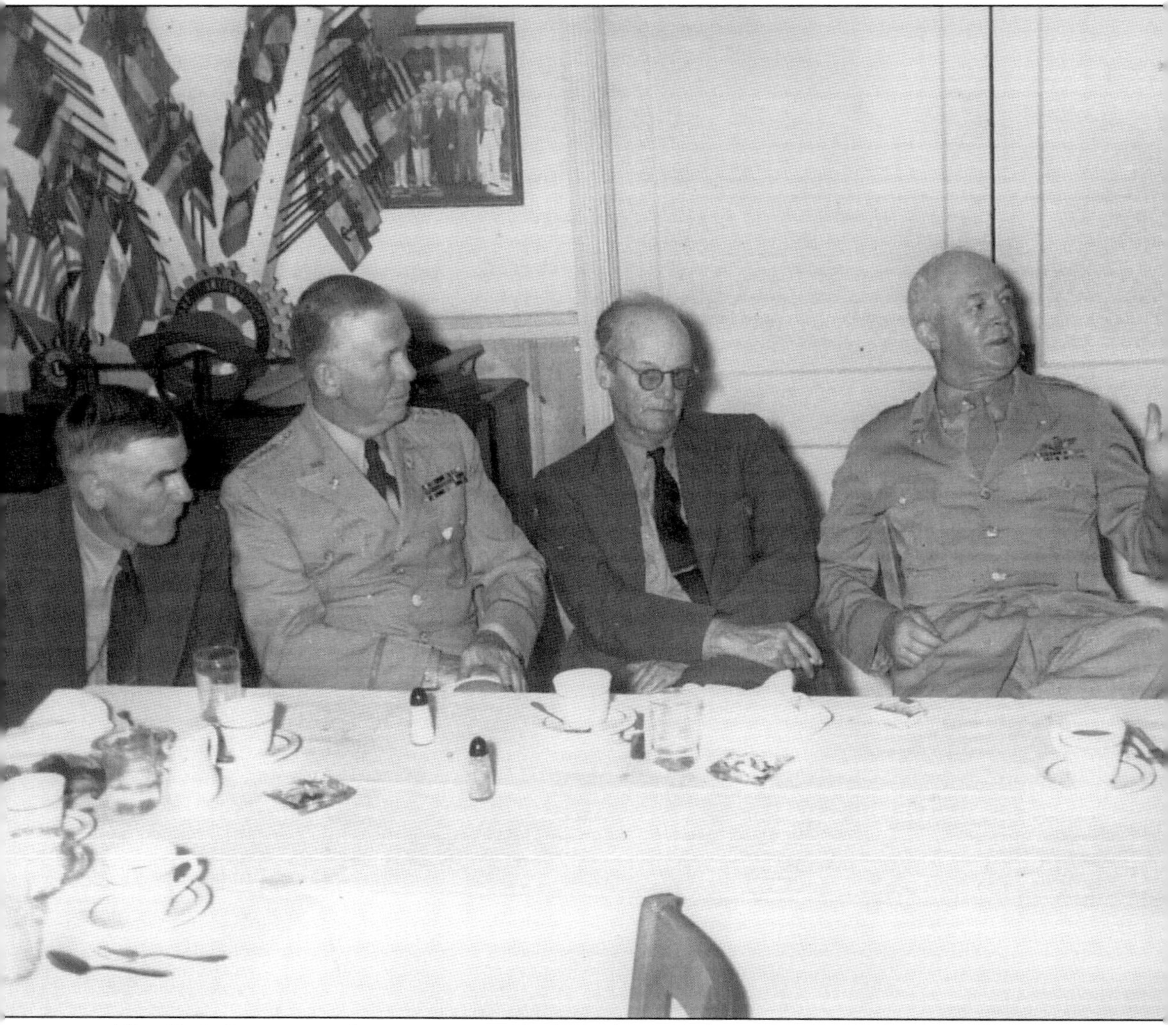

The High Sierra served as a top-secret, high-level respite during World War II. Gen. "Hap" Arnold, chief of the Army Air Corps, had become fondly acquainted with the Inyo and Forest Supervisor Boothe during the CCC days. On August 23, 1944, with the war still raging in both theaters, Arnold and Gen. George C. Marshall, Army chief of staff, flew into Bishop for a weeklong fishing trip with Boothe and two other forest service leaders. The party embarked from Lake Mary and toured south for a week and returned to civilization at Pine Creek. Each morning, they lit a fire with orange smoke to alert their location to a pilot, who would airdrop the latest communiqués about the war. Radio contact was also maintained with the Bishop airfield. Little did they know that they were spotted at Lake Italy by two Japanese-American temporary escapees from the Manzanar War Relocation Center, who saw military uniforms and believed they must be searching for them. After the trip, the generals allowed for this photograph at a restaurant in Bishop. From left to right are Roy Boothe, General Marshall, Regional Forester Bevier Show, and General Arnold. (Courtesy of John Boothe collection.)

Since the early days of the Inyo, some carried their own gear into the high country, because they could not afford to own or hire pack stock, wished to go to places too rugged for animals, or simply wanted to not be tethered to a pack train. Here, Dave McCoy sets out over Piute Pass in the mid-1930s with bedroll, fishing pole, some grub, and clothes. He went alone because he could not talk any of his friends into being their own mules, too. (Courtesy of Dave McCoy collection.)

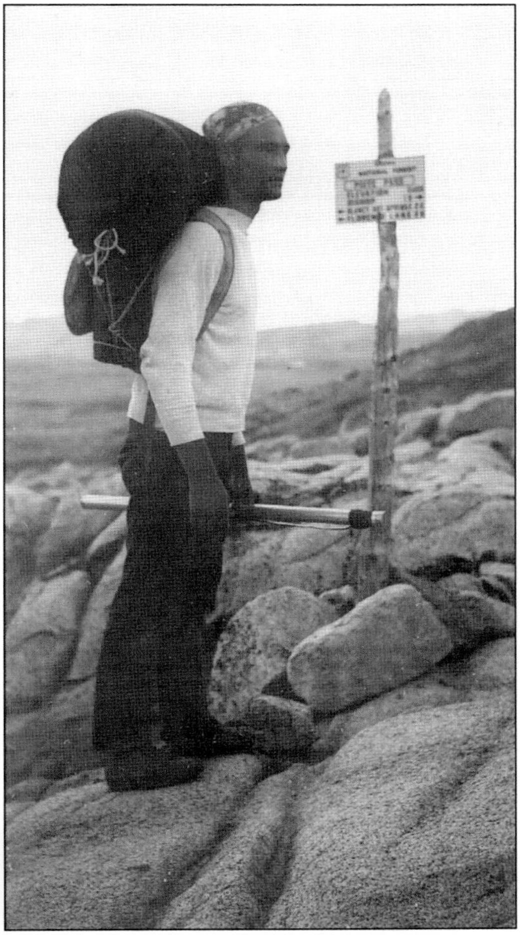

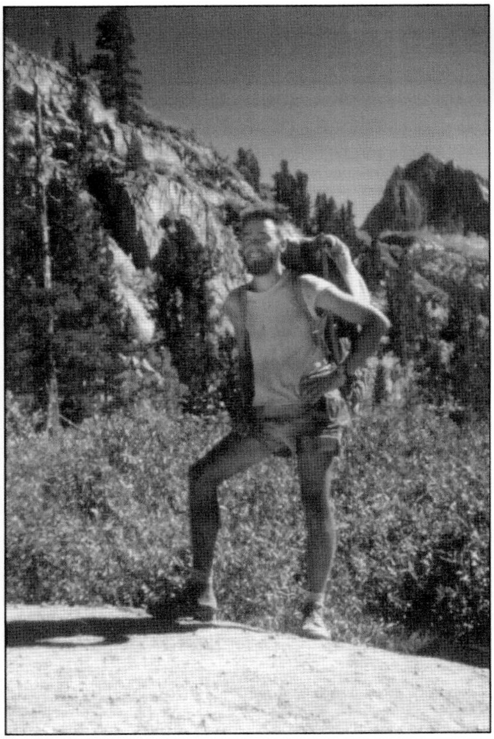

Bishop High School math teacher Ron Smith sets out from South Lake in 1952 for a challenging weeklong backpack trip with his wife, Nancy, and Larry and Laurie Williams. The foursome hiked over Bishop Pass and down the Enchanted Gorge in Kings Canyon National Park. Smith made his own pack frame and strapped a loaded duffel to it. (Courtesy of Ron and Nancy Smith.)

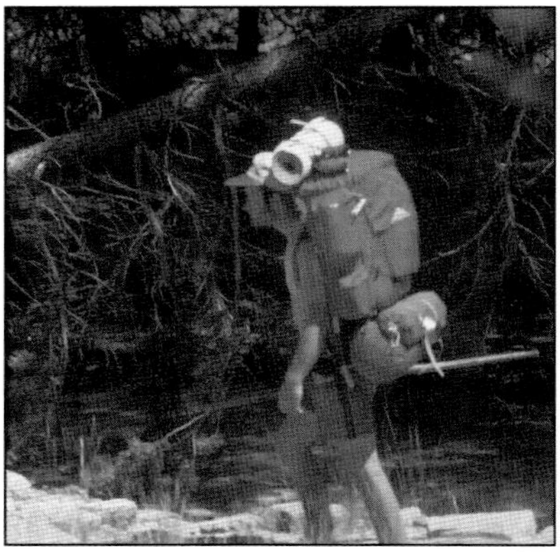

A 1970s backpacker is shown here with the equipment that made carrying one's own burden through the Sierra a reasonable proposition. Kelty frame backpacks, like his, became especially famous after Tom Hornbein used one to climb Mount Everest in 1963. These state-of-the-art backpacks were the standard for some 30 years. The evolving methodology of backpacking came together in the popular book *The Complete Walker*, by Colin Fletcher, published in 1973. A rolled "ensolite" sleeping pad and binoculars are strapped at the top, a down sleeping bag and lightweight tent are strapped to the frame bottom, and inside is clothing and food for a week or so. (Courtesy of Bill Cox.)

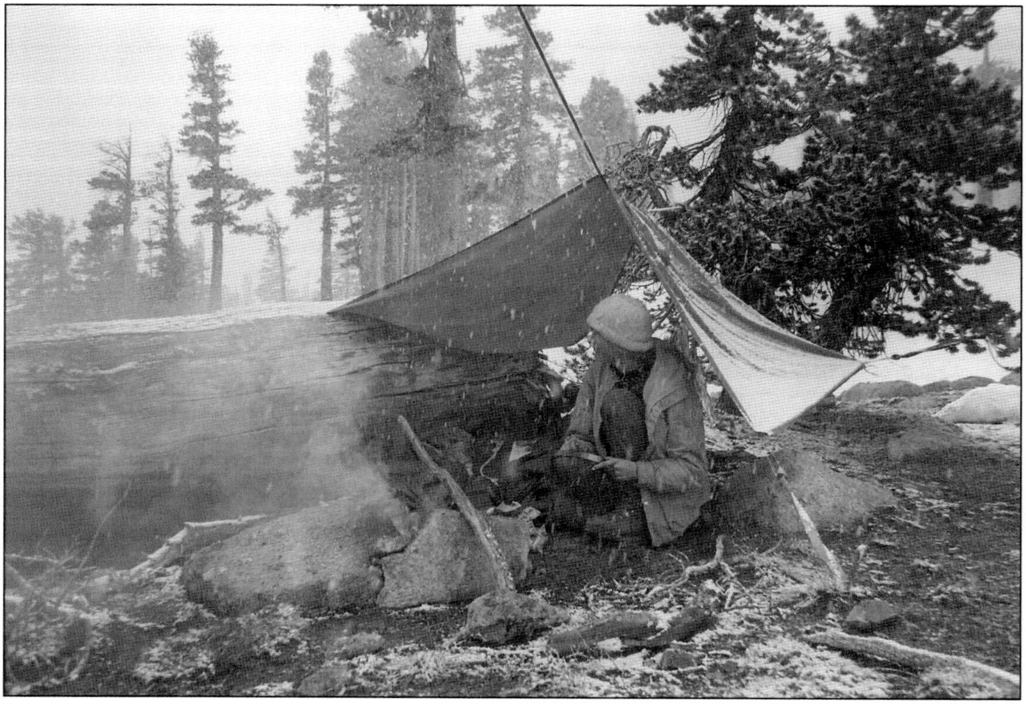

Meredith Wiltsie shelters under a tarp in a June 1975 Sierra snowstorm. Backpackers embraced how carrying less makes for easier hiking, a more intimate experience with the terrain, and the necessity to make-do with less. In the 1970s, a new generation learned to relish the Sierra wilderness in this way. When the weather is fine and things go smoothly, the Sierra seems like a best friend. When things get rough, minimalists have to get along with what they have and/or pay a price. (© Gordon Wiltsie.)

Jay Jensen and Roger Schley relax by a campfire after a day's hike into the Big Pine Lakes in 1974. As high country travel became more popular, backpackers began to prefer camping without fires due to not wanting to leave fire scars or to deplete wood in sparse high country. Also, it became easier to bring along lightweight stoves for cooking. Starting in the 1980s, pressure on fragile high country prompted the Inyo to ban fires in popular basins and, eventually, in most areas above 10,000 feet. (© Gordon Wiltsie.)

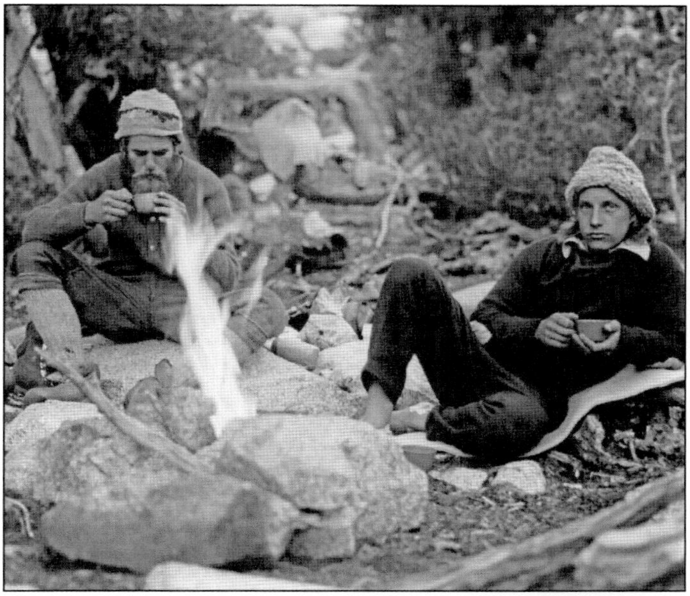

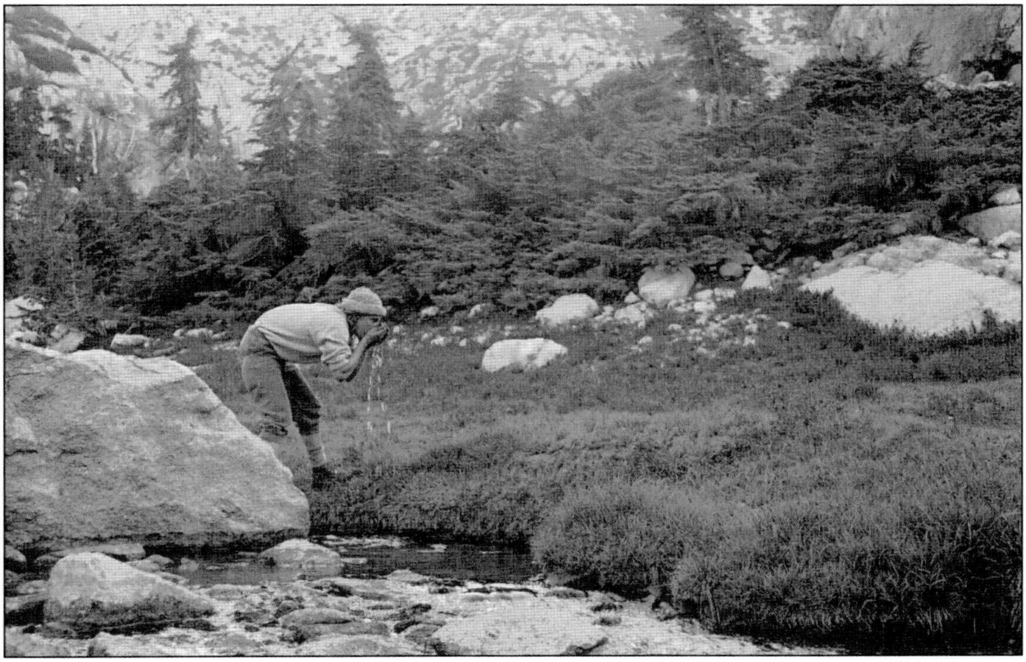

Roger Schley drinks from a High Sierra stream in 1974. Drinking straight from a stream like this was one of the easy joys of backcountry travel, and through the 1970s, most visitors, using some discretion, slurped without a thought. But by the mid-1980s, intestinal disease—colloquially labeled as the parasite "Giardia"—had found a home in some Sierra waters. Also, starting in the early 1970s, the Inyo began to require permits for entering designated wilderness areas. The permit process was an opportunity to inform visitors of various wilderness regulations, hazards, and ethics. These included warning that all water in the mountains should be treated before drinking, and restrictions to protect water quality, namely camping and burying waste well away from streams and lakes. Later, keeping food away from bears became a priority and a legal requirement, too. (© Gordon Wiltsie.)

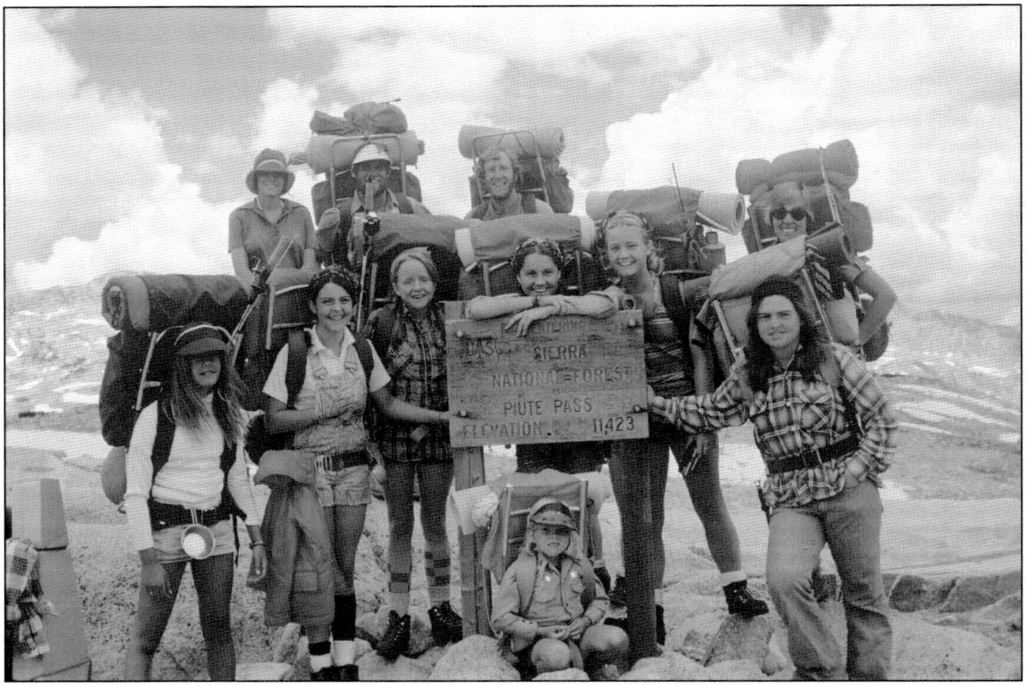

A girls' summer camp group pauses for a photograph at Piute Pass in 1972. Tom and Toni Landis led groups like this on Sierra backpacking tours for several seasons. For decades, groups like the Boy Scouts, YMCA members, university groups, church groups, and others have found that wilderness trips like this help youth grow up together in a vigorous, challenging, restful, and healthy environment. Adult leaders in the back row are, from left to right, Mary Pipersky, Tom Landis, and Bill Cox. Toni Landis is in the center row, far right. (Courtesy of Bill Cox.)

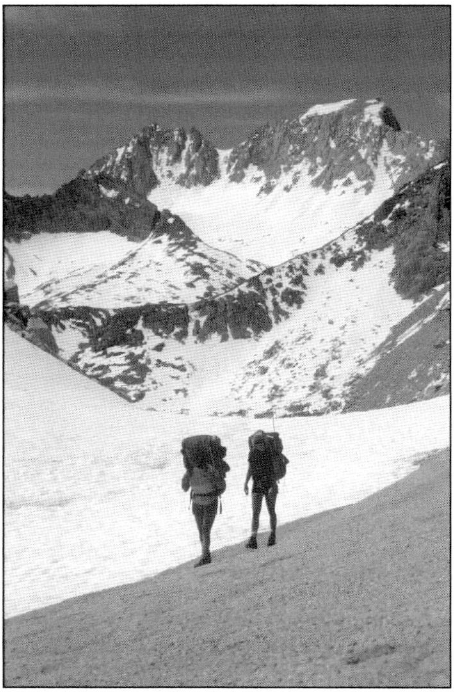

Two backpackers in the mid-1970s cross Mono Pass, beneath Mounts Dade and Abbott. As society evolved, small parties such as this gradually became the norm, and the last big Sierra Club High Trip was in 1972. By 1980, there were likely more people backpacking than traveling with stock. Concern that trails and campsites should remain less trampled pressed the Inyo to establish limits on party size, campfires, and stock use. Also, nearly universal access to private vehicles, general social fragmentation, less leisure time, and worldwide adventure travel put Sierra trips into a different perspective. (Courtesy of Bill Cox.)

Five

From Backwater to Modern World

Management since World War II

In the enthusiastic times after World War II, Americans built more homes and took to the highways to see their "Great Outdoors." Nationally, the Forest Service focused on increasing timber production, but of course, the Inyo had relatively few marketable trees. Inyo Supervisors Slim Davis and then Joe Radel steered the agency into an era where recreation would be not just an assumed amenity but also a primary part of the Forest's mission. They would make sure the Inyo delivered its modest share of wood products, but Radel especially oversaw the development of campgrounds, trails, ski areas, and other recreational facilities. Radel was a follower of the great conservationist Aldo Leopold and he believed that the Inyo should be a place where recreation could foster a deeper connection to land and nature. He often broadcast Leopold's quotes, like "conservation is a state of harmony between man and land."

As Radel neared retirement in 1970, pollution and degradation of many resources prompted the nation to look gravely at its habits, and he was not alone when he told Inyo visitors that America was in "an environmental crisis." In this era, many laws were passed to try to ensure that vital resources were maintained, including the Wilderness Act (1964), the National Environmental Policy Act (1970), the Endangered Species Act (1973), the Clean Water Act (1972), and the National Forest Management Act (1976).

These laws and other policy changes brought substantial benefit for the environment, and they also mandated substantial layers of compliance and planning work for Inyo staff. Supervisors and lower line officers still have discretion, but "significant Forest actions" would now need to gather resource data and solicit public input. This was also the beginning of the end of relative autonomy for permittees, like packers, guides, and ski areas. Into the 1980s, the Forest staff grew with the workload, but then Congress began to restrain and reduce the Inyo's budget. With pressure to do more with less, the staff began to spend much less time in the field. However much the struggle with paperwork, the staff and the public take heart that, even though the Inyo sees over four million visitors a year, it remains a beautiful, largely intact natural landscape.

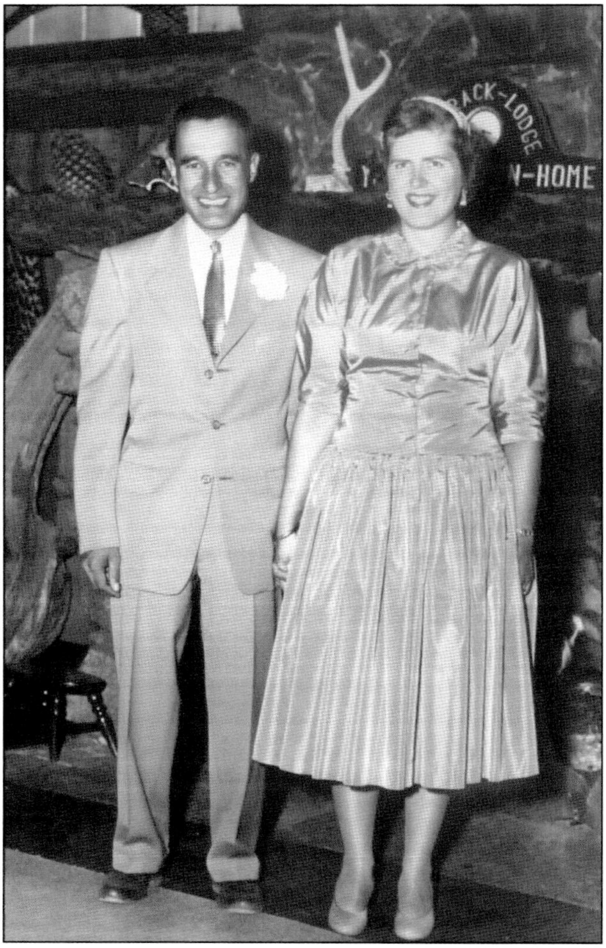

Edwin "Rocky" Rockwell was a World War II veteran from Vermont who worked for the Inyo from the early 1950s to 1980. The September 1955 image at left shows Rockwell with his bride, Tish, at their wedding reception at Tamarack Lodge. He became one of the Inyo's first "winter sports specialists," and as Mammoth Mountain Ski Area developed, they were among the first Inyo staff to live through inhospitable Mammoth winters at the Mammoth Ranger Station (below), a rustic place with no indoor bathroom. The ski area hired Tish to cook for employees, and she later became a clerk for the Inyo. Rocky was also a firefighter and a media liaison. In the 1960s, he began to coordinate and appraise land swaps that conveyed National Forest land around Mammoth into real estate in exchange for private inholdings in other areas of the Inyo. The Rockwells founded the Eastern Sierra Interpretive Association in 1970 to help educate people about public lands. (Left, courtesy of Rocky and Tish Rockwell collection; below, courtesy of INF.)

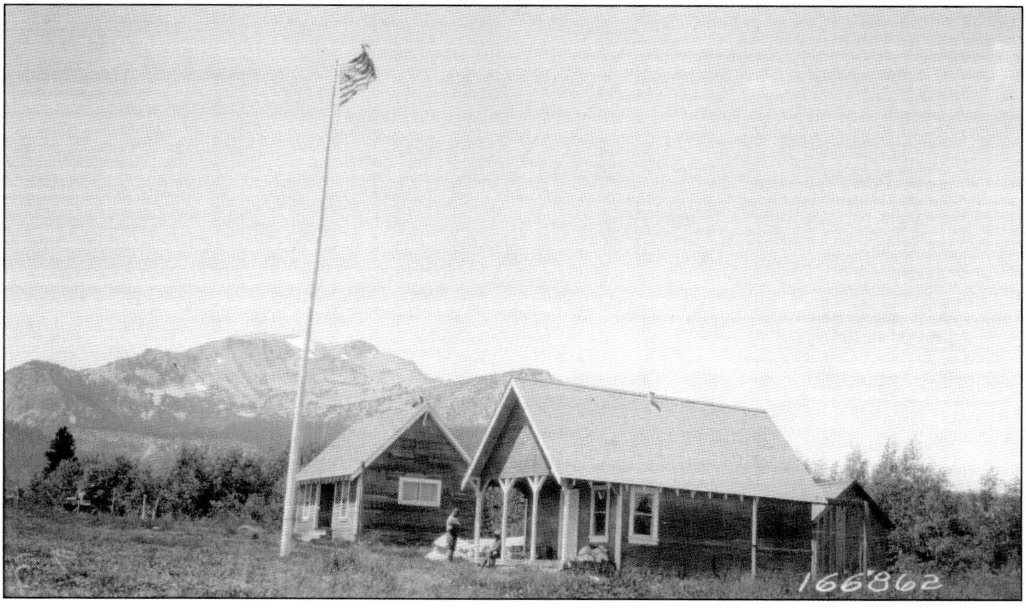

Since 1915, Congress has mandated the Forest Service to designate lots where members of public are, for a fee, permitted to build and use a cabin for part-time, recreational residence. On the Inyo, most permits were issued from the 1920s to the 1950s. Since then, many cabins in high-value recreation areas have been moved to sites with less public interest. Today, the Inyo manages 22 tracts with 350 cabins on 20-year leases but has no plans to expand the program. These cabins are part of the Falls Tract, near Lake Mamie, in the Mammoth Lakes Basin. (Courtesy of AS.)

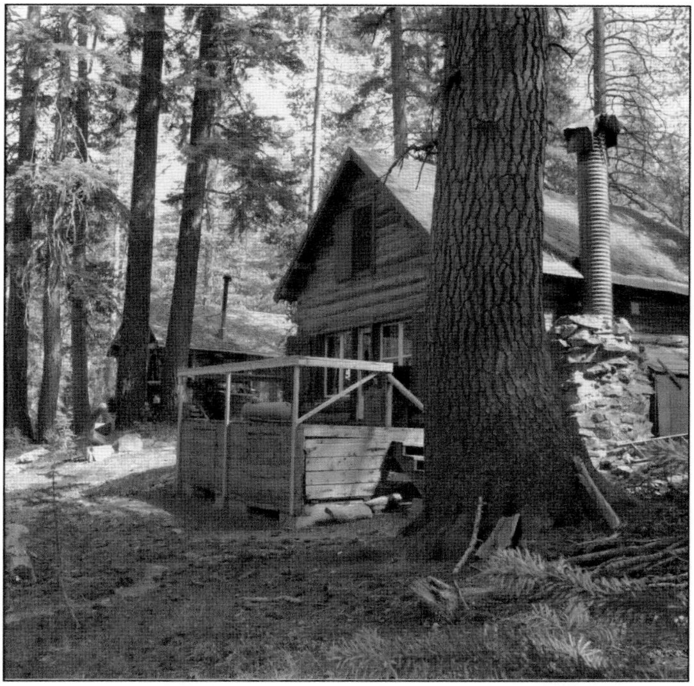

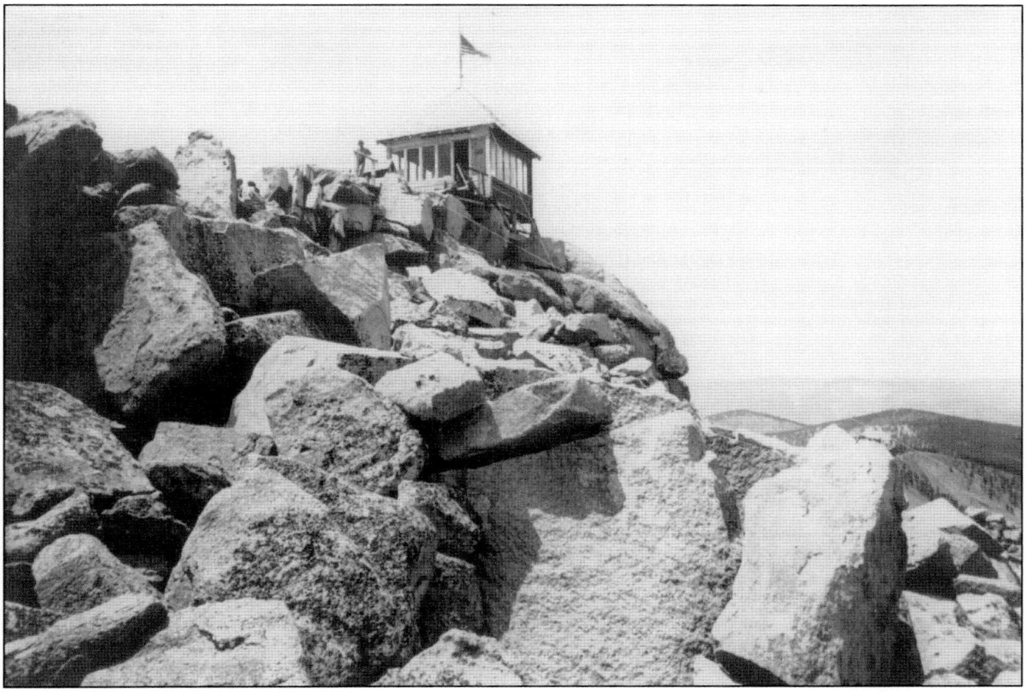

For most of its existence, the US Forest Service's core mission included protecting forests from fire. Relative to other national forests, the Inyo has fewer trees and less fuel, but brush and forest fires do occur, and managing fire is a major agency task. The Kern Plateau, in particular, has plenty of forest, and long ago, the agency erected a lookout atop Kern Peak—a perch with a fine view over the whole region. The Inyo's only other lookout is at Bald Mountain, overlooking the northern span of the forest from the Glass Mountains. (Courtesy of INF.)

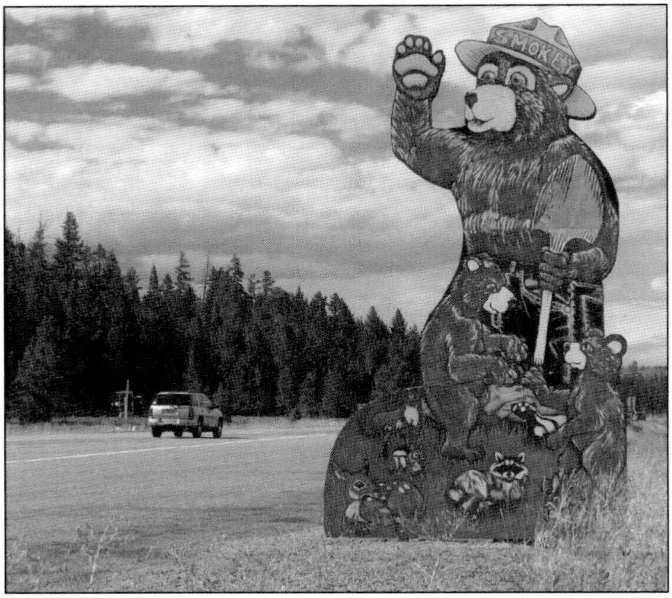

In 1944, the Forest Service and the Ad Council came up with Smokey Bear to convey the message "to end all forest fires." Today, the agency recognizes that extensive fire suppression has let fuels accumulate so much that wildfires have become catastrophic more often. Smokey's message to the public is still valid, however, as carelessly started fires are still a serious hazard. This 15-foot-high statue of Smokey was crafted by firefighter Dick Dahlgren in the mid-1950s and is erected every spring along Highway 395 north of Mammoth. (Courtesy of AS.)

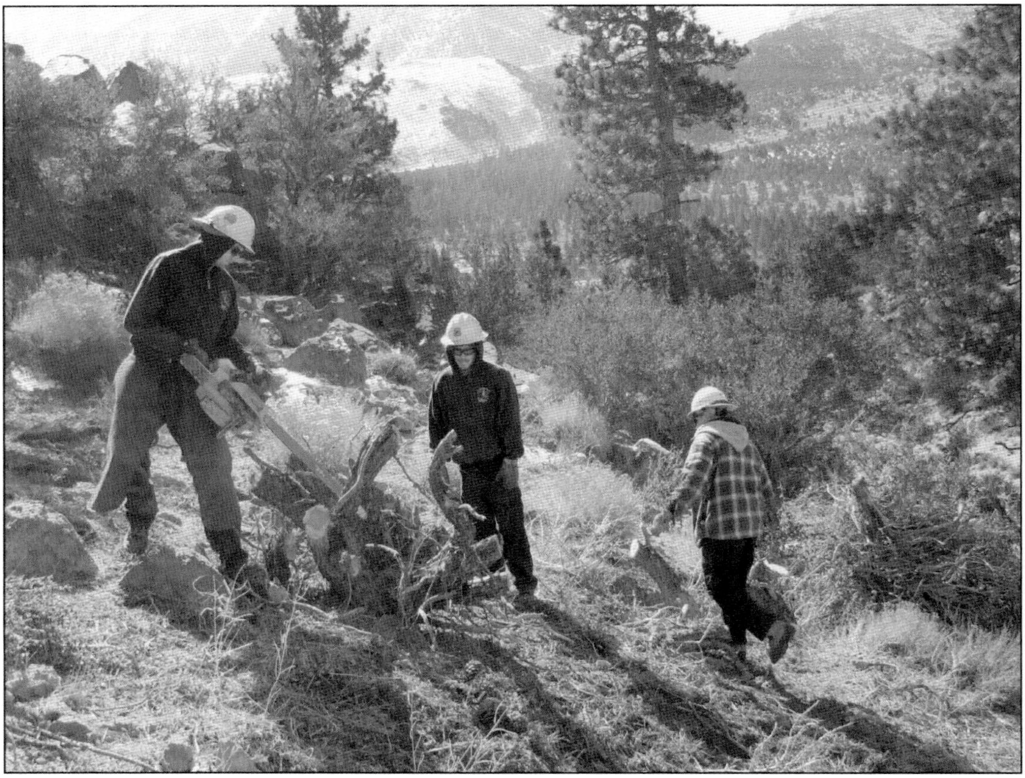

The Forest Service today recognizes that safe and productive forests need occasional fires. On the Inyo and elsewhere, the agency conducts prescribed burns during safe conditions. To reduce the fire hazard near structures, crews remove accumulated brush and thins trees. Homeowners with forest surroundings are required to create "defensible space," while agency crews reduce the fuels around those communities. Here, an Inyo crew thins brush and Jeffrey pines near Toms Place. (Courtesy of AS.)

A plane drops fire retardant on the Birch fire, a major blaze near Sherwin Summit in 2002. Extinguishing large fires like this is still a major responsibility for the Inyo. All stops were pulled to put out this and some other recent eruptive blazes that have threatened communities. Into the 1970s, all able-bodied Forest Service employees were notified that they could be called on fire duty anywhere in California. (Courtesy of AS.)

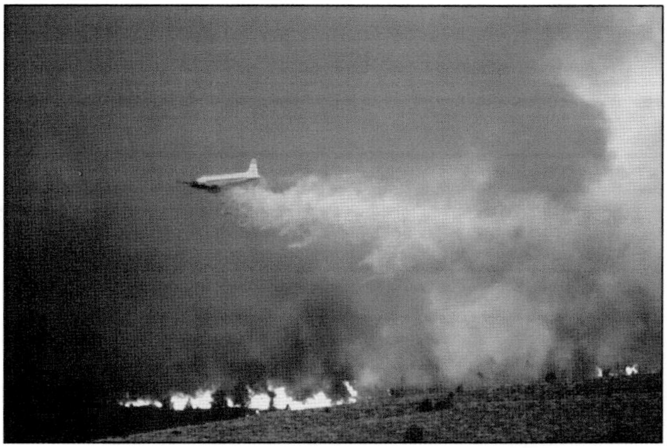

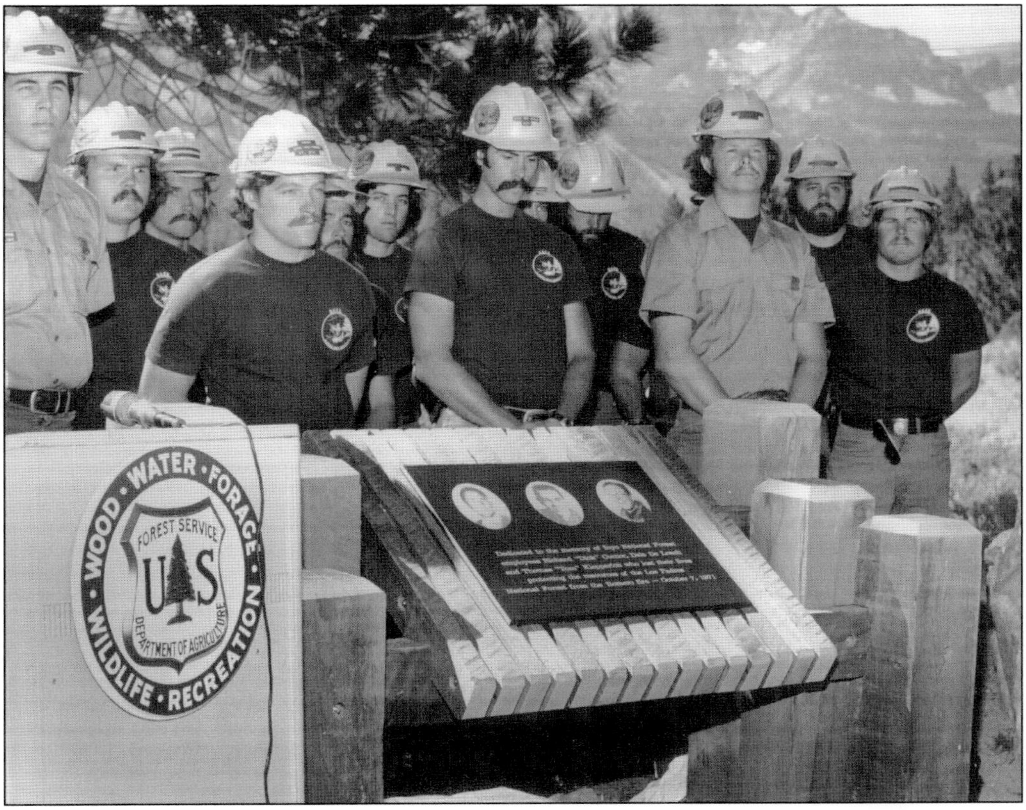

The Inyo "Hotshots," a crew of wildland-trained firefighters, surround a memorial erected in 1975 on Oh! Ridge above June Lake to honor four men who were killed in action. Firefighters Dick Cumor, Dale De Loach, Tom Klepperich, and contractor James Monreau died fighting the Romero fire in the Los Padres National Forest on October 7, 1971. From left to right are foreman Rick McCool, Doug Morrison, Michael Milligan, Dan Whitmore, Al Hunter, Tony Brunwin, Ken McCool, unidentified, Greg Clausen, foreman Lance Hiester, Mike Nelson, and Steve Burns. (Curtis Phillips photograph, courtesy of INF.)

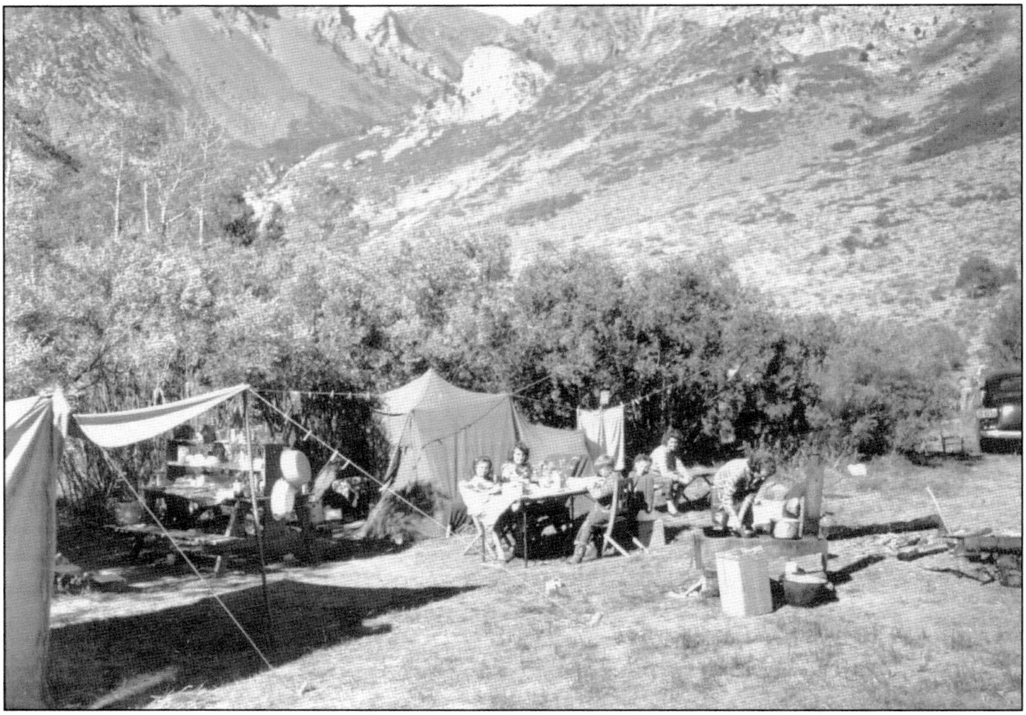

As the Inyo understood that it needed both to foster visitation and manage the visiting public, they developed a variety of recreation facilities, including improved roads and designated campgrounds. Above, a family camps at the new McGee Creek campground in the 1940s. Camping was becoming popular enough by then that merchants like Sears and Roebuck carried tents and agencies like the Inyo were designating campgrounds with tables and restrooms. Below, the staff plans the Convict Lake campground in about 1958. Shown below are Inyo recreation officer Lloyd Hayes (below the others) and Inyo landscape architect Wayne Iverson (left); Mammoth district ranger Bill Murphy (center); and Earl Bachman, regional recreation officer. (Courtesy of INF.)

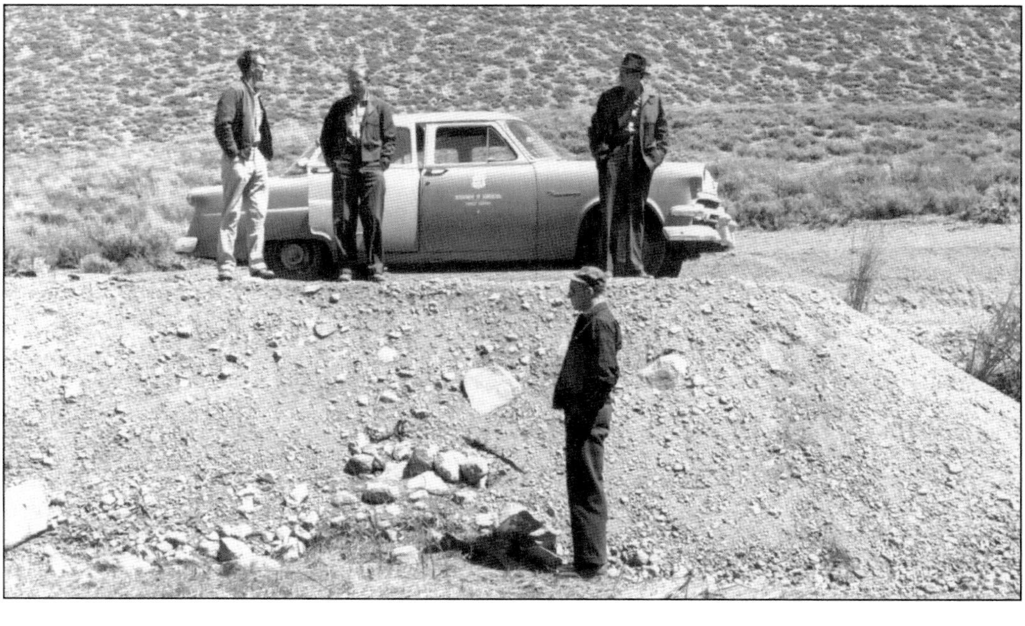

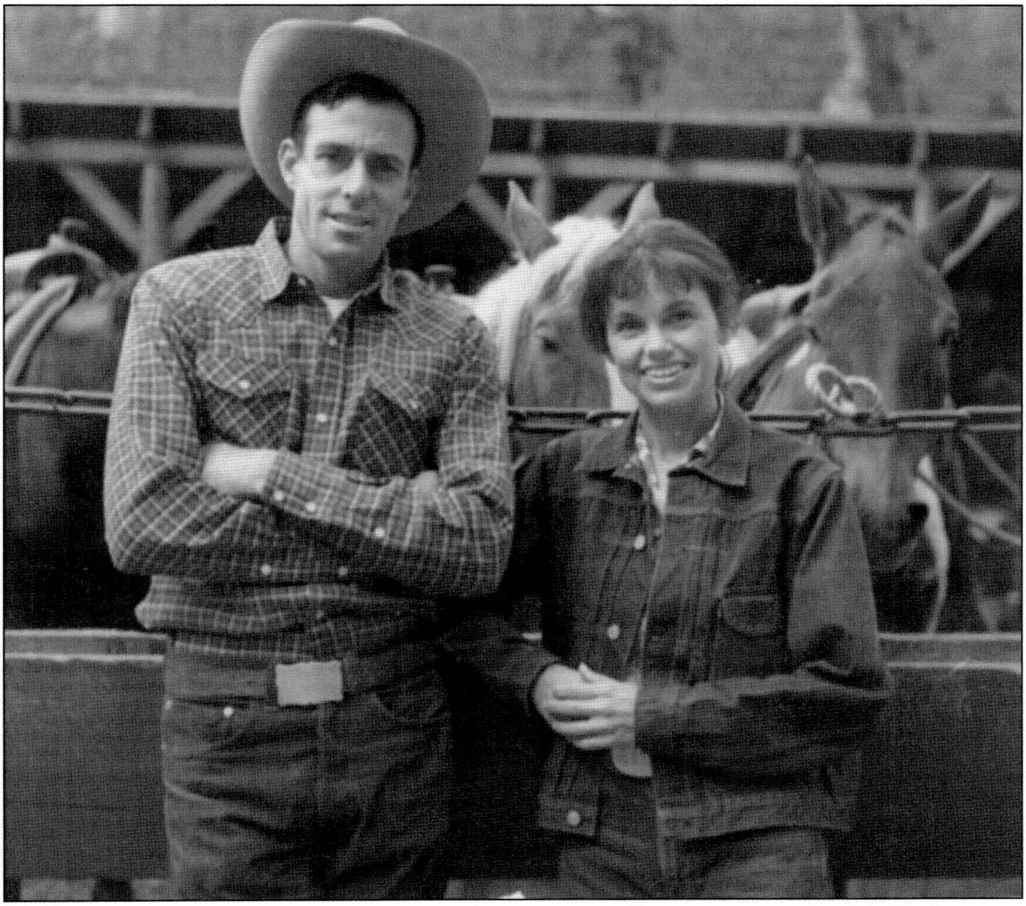

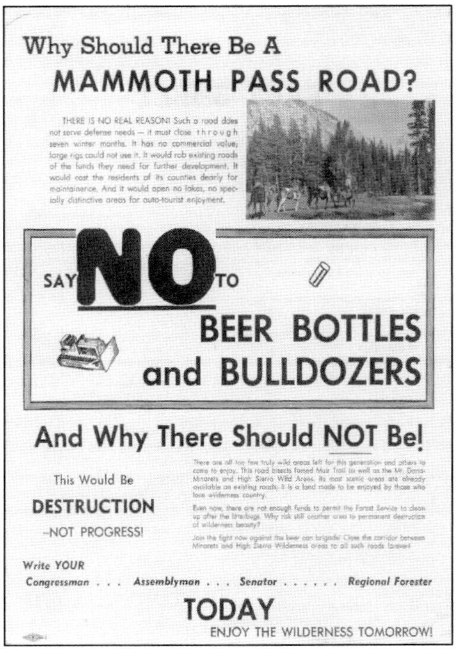

Lou and Marye Roeser, owners of Mammoth Lakes Pack Outfit, were among the citizens who blocked a proposed four-lane highway across the Sierra from Fresno over Minaret Summit. When it designated the John Muir and Minarets Wilderness Areas in 1964, Congress excluded a corridor up the San Joaquin River Canyon for this plan. Pressure from San Joaquin Valley leaders to build this "Road to America" was intense, but the Roesers, Genny Smith, Bob Schotz, Bob Tanner, Doug Kittredge, and other Mammoth residents formed an opposition group, the Mono County Resources Committee. They argued that the backcountry was more valuable as wilderness than as a highway thoroughfare and that the costs of construction and winter maintenance would be far greater than advertised. On April 18, 1967, Lou and others successfully lobbied on the floor of the California legislature, and the plan was voted down. (Above, courtesy of Lou and Marye Roeser collection; right, courtesy of Devils Postpile National Monument.)

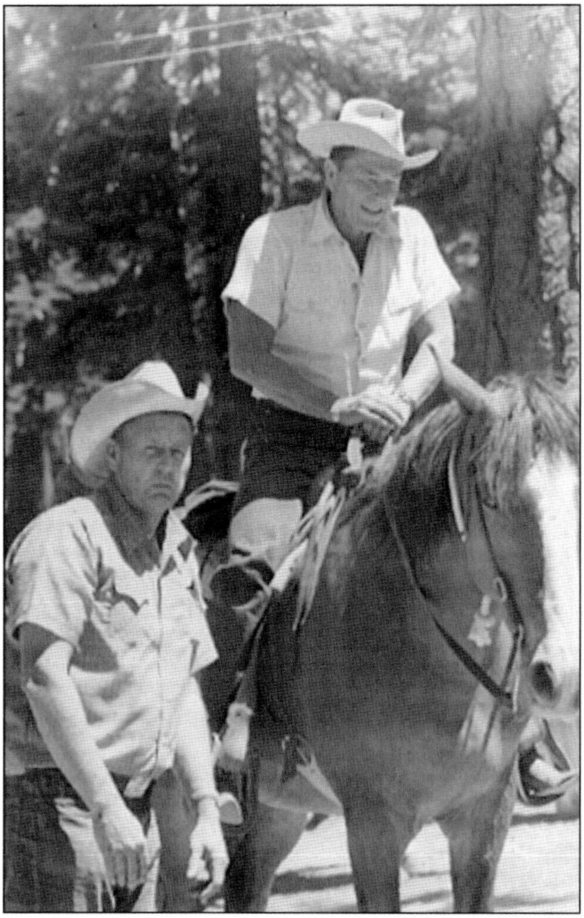

After California rejected the trans-Sierra highway plan, proponents persuaded the US Department of Agriculture to fund the Forest Service to build it. Ike Livermore, Governor Reagan's resource secretary, was opposed. He took Forest Service officials on a wilderness pack trip to show them the wilderness that would be harmed. Ironically, Ansel Adams and the Sierra Club were strategically supportive of the highway, arguing that it would take pressure off of other trans-Sierra highway proposals. Governor Reagan wrote to President Nixon asking for the removal of federal funding. On June 28, 1972, Reagan rode out from Reds meadow to announce that the president had withdrawn funding. Left, the governor saddles up next to Reds Meadow Pack Station owner Bob Tanner. Below, the governor lunches with packers Arch Mahan (left) and Herb London. In 1984, Congress passed new legislation closing the proposed highway corridor and including it in an expanded Wilderness Area named for Ansel Adams. (Courtesy of Lou and Marye Roeser collection.)

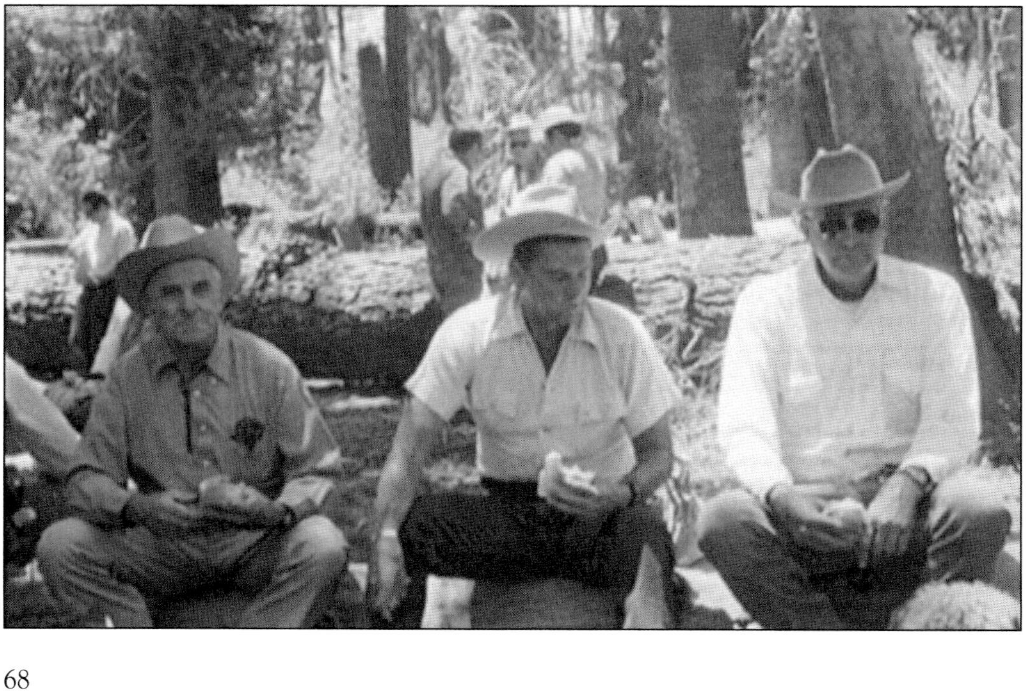

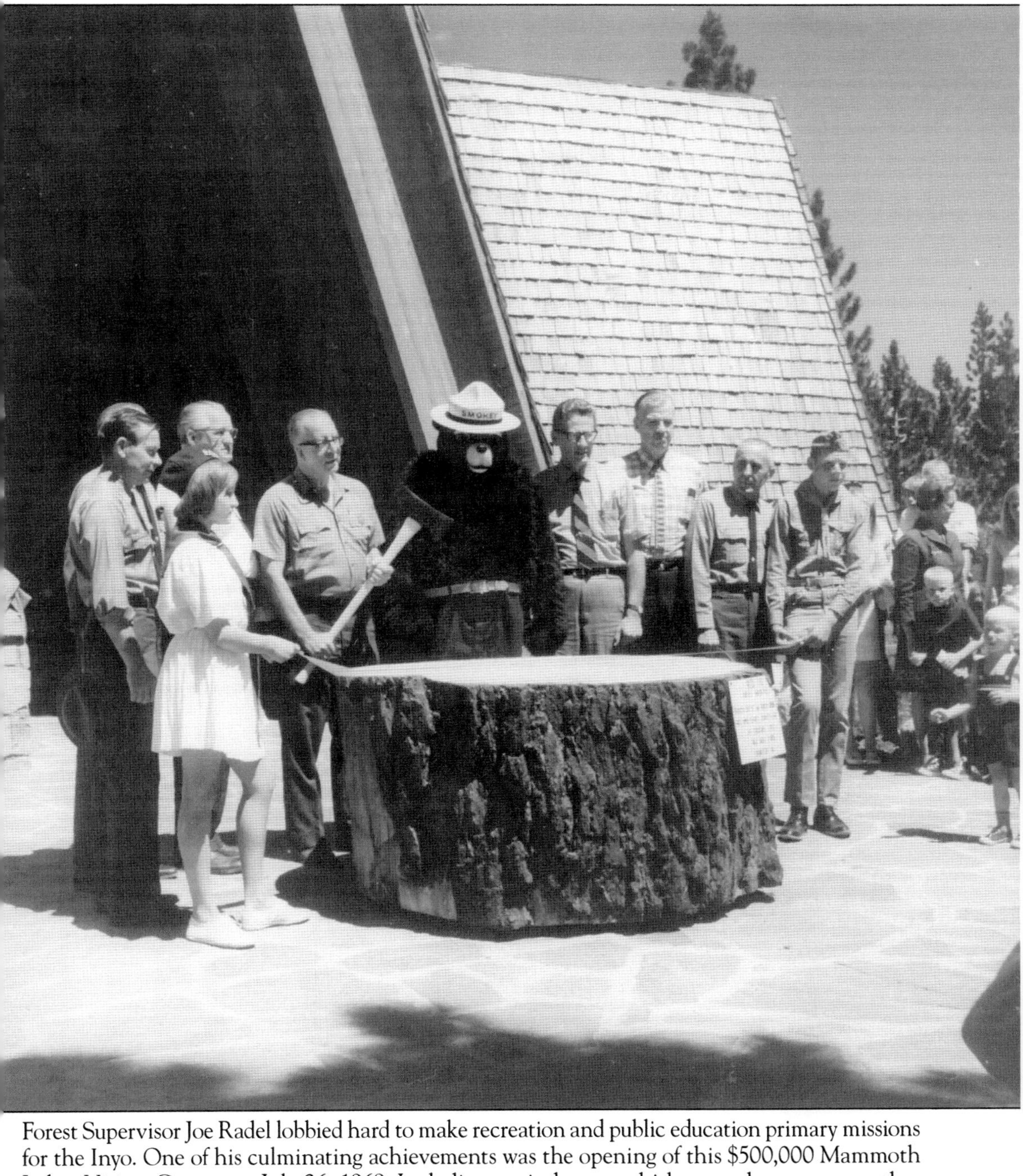

Forest Supervisor Joe Radel lobbied hard to make recreation and public education primary missions for the Inyo. One of his culminating achievements was the opening of this $500,000 Mammoth Lakes Visitor Center on July 26, 1969. Including an indoor amphitheater, there was no other facility like it in the Forest Service. It was built partly on the idea that the agency should expand its mandate and serve the public similarly to the National Park Service. From left to right are an unidentified Girl Scout; Mammoth district ranger Bill Murphy; Regional Forester Jack Deinema (hidden); Congressman Harold "Bizz" Johnson; Chief Ed Cliff (with axe); Dick Trambley as Smokey; Assistant Secretary of Agriculture E.J. "Fritz" Behrens; California resource secretary Ike Livermore; Supervisor Radel; and an unidentified Boy Scout. (Courtesy of INF.)

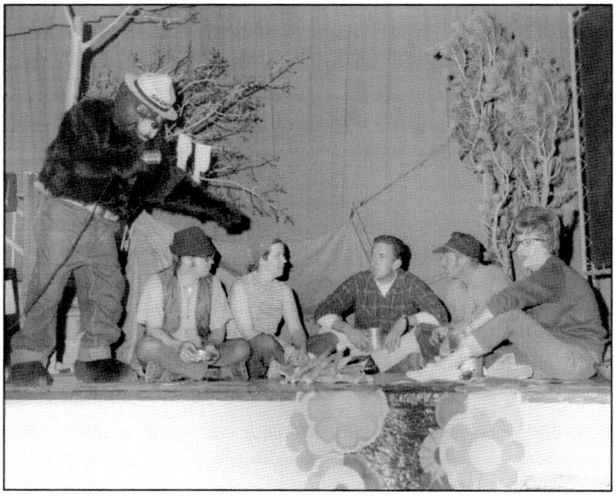

In March 1971, Smokey Bear is shown acting in a public education skit, "The Careless Camper." At the new amphitheater in Mammoth, the audience learned not only how to be careful with fire but also how to camp with a minimal impact on vegetation and water. From left to right are Jerry Cimino as Smokey; Dick Harris, Inyo naturalist; Connie Mitchell, wife of local Sierra Club leader Roger Mitchell; Dick Cumor, a wilderness ranger who died firefighting the following summer; Bill Harvey, Inyo staff; and Helen Jones. (Courtesy of INF.)

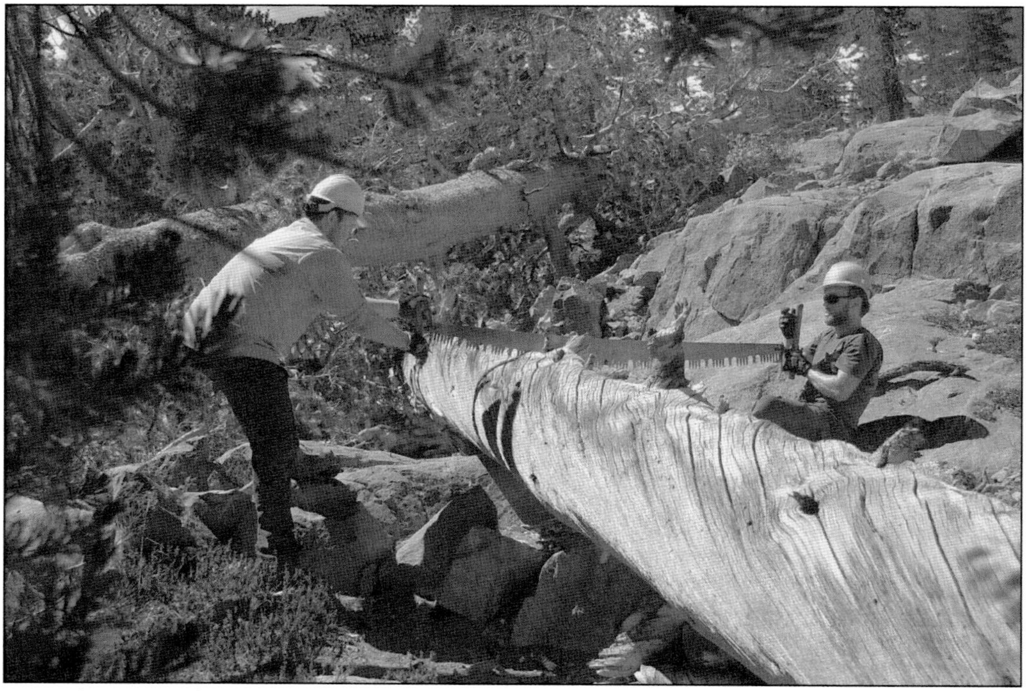

Since the 1980s, the Inyo has struggled with reduced budgets from Congress even as compliance requirements and recreation demands have increased. The Forest has turned to volunteer service help, often to Friends of the Inyo (FOI), a nonprofit founded in 1986. In 2011, Andrew Schurr (left) of FOI and volunteer Cory Meza are shown clearing deadfall from the trail to Shadow Lake. In wilderness areas like this, they use not a chainsaw but a traditional crosscut saw. (Courtesy of Jo Bacon.)

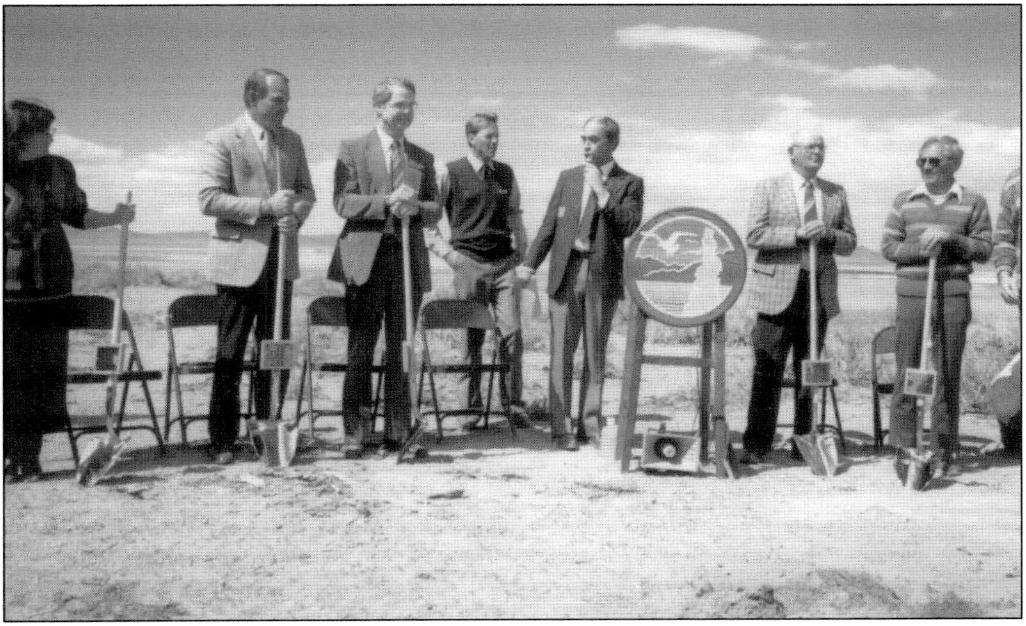

Increasingly confident in its role as environmental steward, the Inyo brought its weight to the Mono Lake issue. The lake was diminishing from water diversions to Los Angeles, and the nonprofit Mono Lake Committee was campaigning to force the city to allow flow into it. In 1983, Congress designated the basin as the nation's first National Forest Scenic Area. Through land exchanges, the Inyo also acquired more land around the lake. In 1992, it opened a magnificent visitor center, on top of what had once been a landfill for Lee Vining. Here, Forest Supervisor Dennis Martin inaugurates ground breaking for it. From left to right are Martha Davis, Mono Lake Committee executive director; Duane Buccholz, district engineer for Los Angeles's DWP; Dick Benjamin, Forest Service recreation director; Bill Bramlette, Mono Lake district ranger; Supervisor Martin; Don Banta, prominent Lee Vining businessman; and Don Rake, Mono County supervisor. Below, the visitor center is seen overlooking the lake. (Above, courtesy of INF; below, courtesy of AS.)

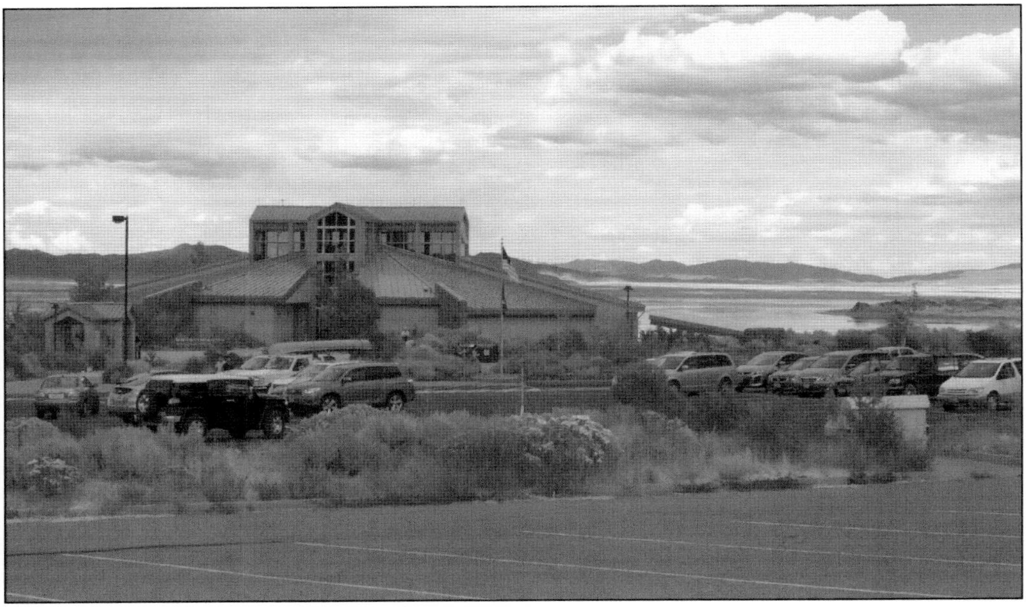

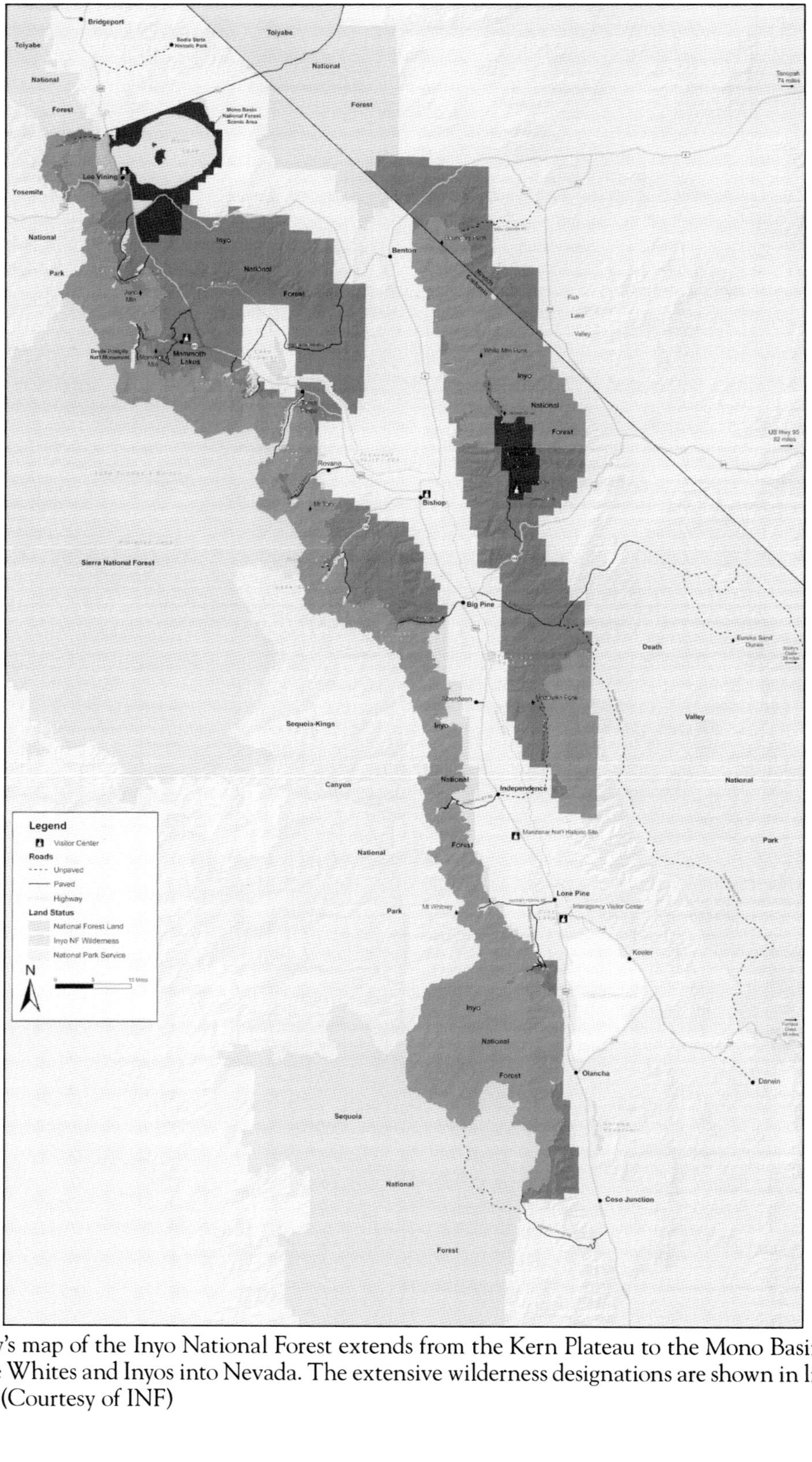

Today's map of the Inyo National Forest extends from the Kern Plateau to the Mono Basin and in the Whites and Inyos into Nevada. The extensive wilderness designations are shown in lighter tone. (Courtesy of INF)

Six

Wings over the Sierra

Skiing

For ages, winter snow made the Sierra's rugged eastern front an utterly inaccessible netherworld. But as European immigrants introduced "Norwegian snowshoes" to California, people learned to take to the Sierra's white season with relish. They found the snowpack to be often skier friendly, and winter's huddled intimidation transformed into enthusiasm that on snow one's feet could be like wings.

Some Mammoth miners started skiing during the 1870s boom, probably when some of the skiing miners from Plumas County sifted south. Starting in the 1880s, Mono Basin pioneer Louis DeChambeau started making skis, while Norwegian miner Erik Erickson introduced technique. But in general, skiing remained obscure. The modern ski era began in the 1920s, particularly with European immigrants. By the 1930s, clubs and groups of skiers were developing rope tows on slopes throughout the Inyo, from the Whitney area to the Mono Basin.

In the early 1950s, the Inyo gave the go-ahead to an ambitious skier named Dave McCoy to start building a more substantial ski operation at Mammoth Mountain. Situated in the gunsight lee of relatively low Mammoth Pass, Mammoth's volcanic slopes collect probably more snow than any other mountain in the Eastern Sierra. With Forest Service endorsement, McCoy's drive, and the Los Angeles market, Mammoth Mountain Ski Area expanded from a cottage industry into one of the largest and most popular ski facilities in the Western Hemisphere, with skier visits accounting for about a quarter of the entire visitation on the Inyo. "The Mountain" became like a gravitational field, dominating not only winter activity but also making the hamlet of Mammoth Lakes into the investment and population center of Mono County.

While tens of thousands of weekend skiers drove up Highway 395 to Mammoth, smaller numbers of "backcountry" skiers continued to find adventure and solitude on the multitude of other facets of the Sierra. Today, even as the joy of skiing can seem small compared to the push for resort development, tens of thousands of skiers—and now snowboarders—still cannot wait for when snow blankets the Sierra.

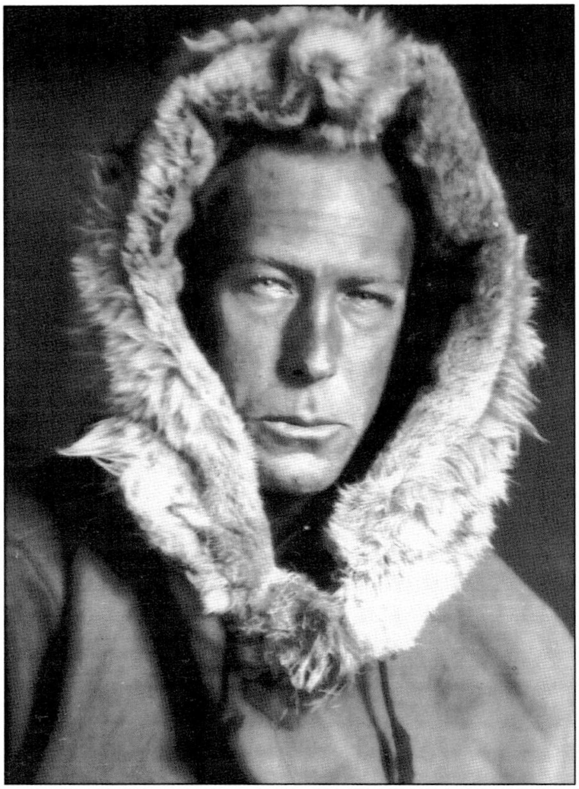

Orland Bartholomew was a pioneer for skiing the higher Sierra. A snow surveyor from Visalia, Bartholomew realized that the High Sierra in winter was a polar-like wilderness waiting for its first exploration. In 1928, he set out 11 food caches, intending to ski from Mount Whitney to Yosemite. On Christmas Day, he embarked over Cottonwood Pass on homemade skis, with a kit including a down robe and a double-bit axe. Within two weeks, he made the first winter ascents of Mounts Langley and Whitney. For three months, Bartholomew continued north, camping through severe winter storms with fires and a tarp. On April 3, he arrived in Yosemite Valley, completing one of the great adventures in the history of the American West. Bartholomew is pictured at left, and below are remnants from one of his caches, discovered and brought down from near Kearsarge Pass. (Left, courtesy of Phil Bartholomew; below, courtesy of AS.)

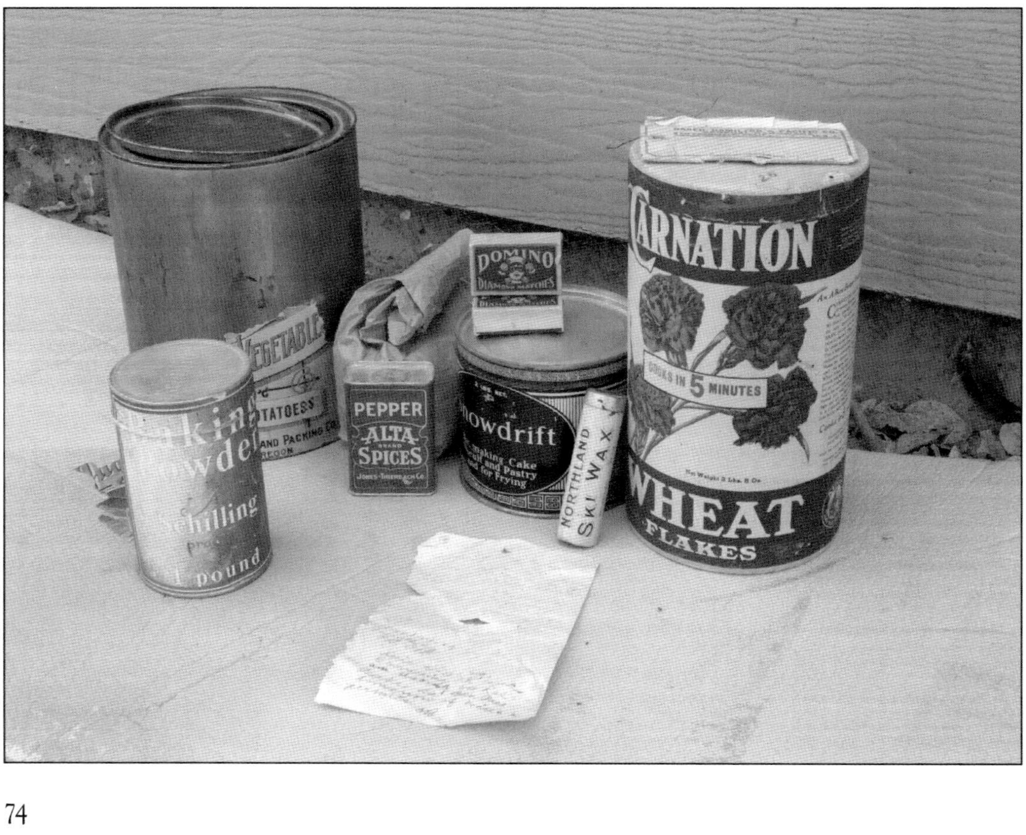

By about 1930, Los Angeles started hiring skiers to measure the snowpack to predict annual runoff. At right, employee Dave McCoy measures the snow with a Mount Rose snow tube. The sectioned, hollow tube is rammed through the snow to the ground. The surveyor then can measure the snow depth and weigh the filled tube to estimate snowpack density. Every year, the same courses are measured at the same times, so over years, the runoff estimates have become reasonably accurate. Today, the California Department of Water Resources, in conjunction with the San Joaquin Valley agricultural community, still hires ski surveyors to use the same methods. (Courtesy of Dave McCoy collection.)

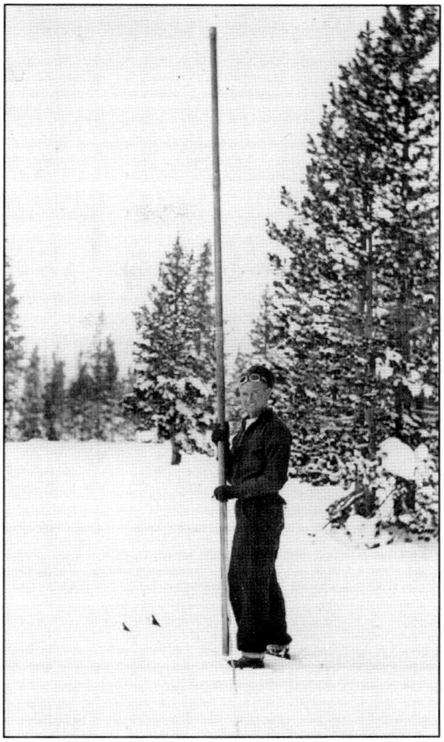

The Swiss-American H.F. Rey family pioneered ski vacationing to Mammoth, starting in 1927. They frequently took tours from "Old Mammoth" up to the lakes basin and along Sherwin Ridge, as shown here. They bought a cabin and rigged a rope tow nearby. They made an Easter tradition of skiing to the top of Mammoth Mountain (then called Mount Mammoth) and enjoying the long descent back. (Courtesy of Rey family collection.)

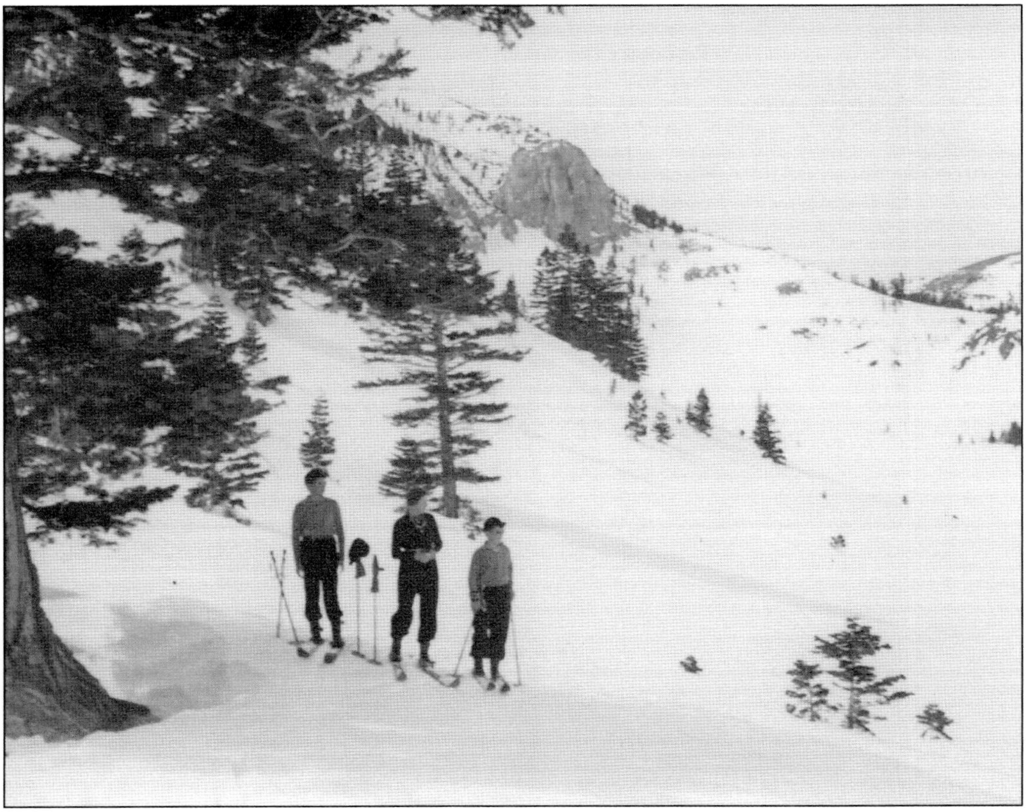

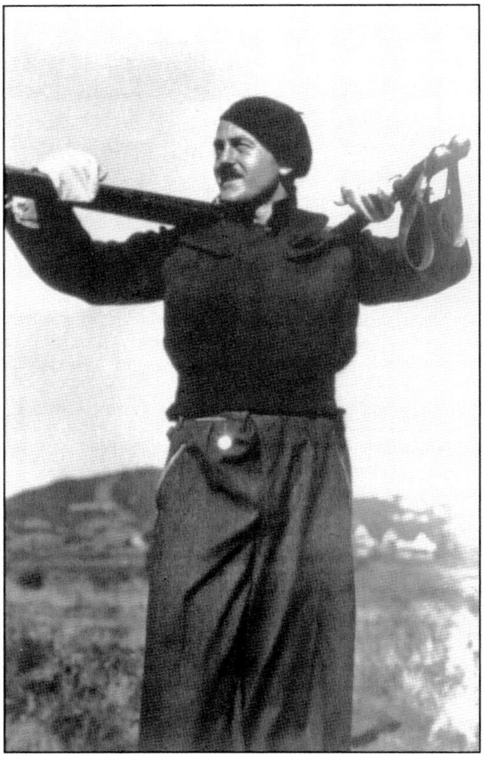

In the 1930s, German and Austrian immigrants introduced skiing to many Californians. Two, in particular, were Walter Mosauer (left) and Wolfgang Lert (below). Mosauer was a zoologist at the University of California, Los Angeles, and, in 1934, he founded the Ski Mountaineers as a section of the Sierra Club. Mosauer and Lert drove north to ski in the Sierra many times, often bringing Sierra Club mountaineers with them. Some, like Glen Dawson and, especially, his brother Muir, became proficient. Mosauer unfortunately died from an illness on a zoology excursion in Mexico in 1937, but Lert has continued to ski into the 21st century. (Both images courtesy of Ingrid P. Wicken.)

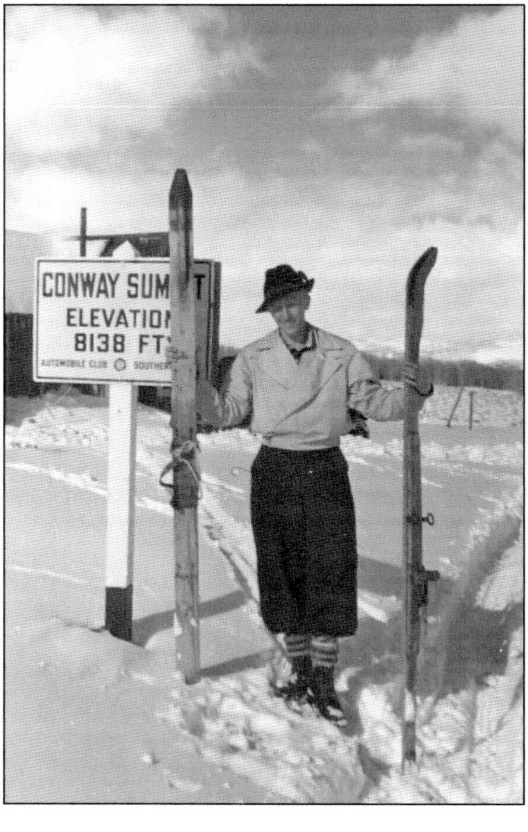

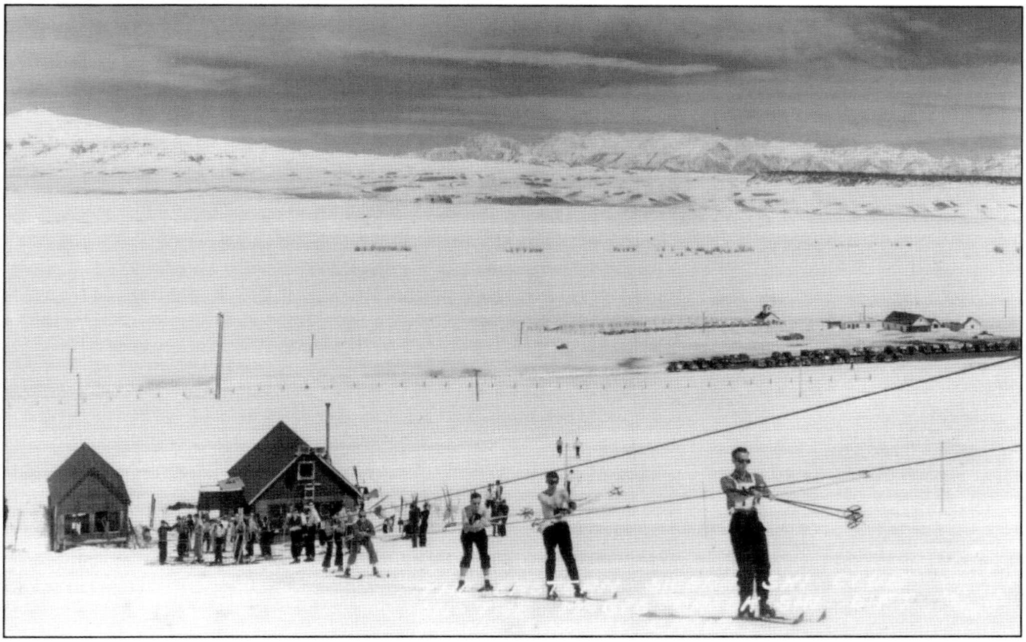

Skiers with mechanical and engineering backgrounds began to devise methods to get themselves dragged up a slope to then enjoy the *schuss* down. Most common were rope tows driven by a jacked-up vehicle. Such arrangements were rigged at Whitney Portal, Onion Valley, Glacier Lodge, Little Round Valley, Mammoth, Crestview, Fern Creek, and elsewhere. The best attended was at McGee Mountain, a convenient slope by the highway above Long Valley. Jack Northrup of aircraft fame designed a tow with an overhead tower system and called it the "Upski," pictured here in 1938. His friend Corty Hill and others built and installed it here. (Courtesy of Burton Frasher collection.)

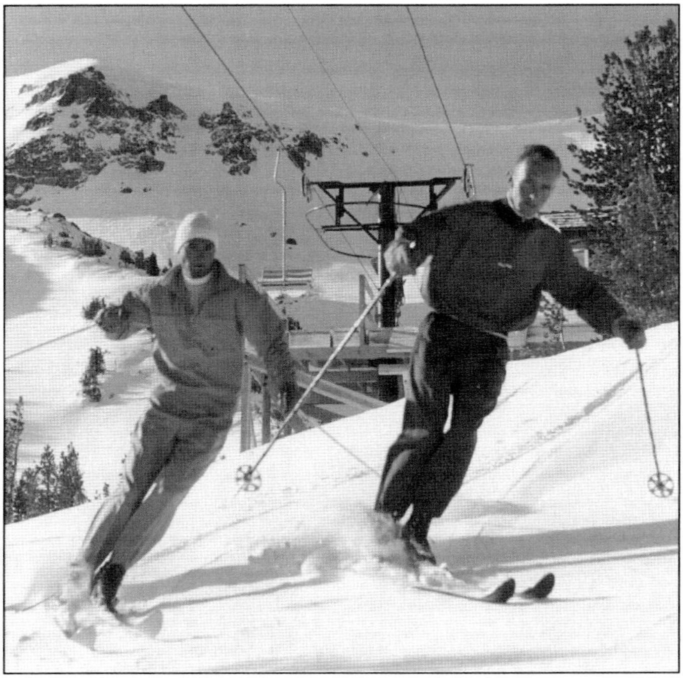

In 1953, Forest Supervisor Davis gave Dave McCoy the go-ahead to develop a ski area at Mammoth Mountain. By the late 1950s, McCoy and partners had built two long chair lifts extending up Mammoth Mountain. Here, Pete Mead and Bobby Cooper enjoy the terrain on newly built Chair No. 2 in about 1958. A Vermonter, Mead was the brother of Andrea Mead Lawrence, America's greatest Olympic ski racer. She later became one of Mammoth's most prominent and visionary citizens. (Courtesy of H.F. Haemisegger.)

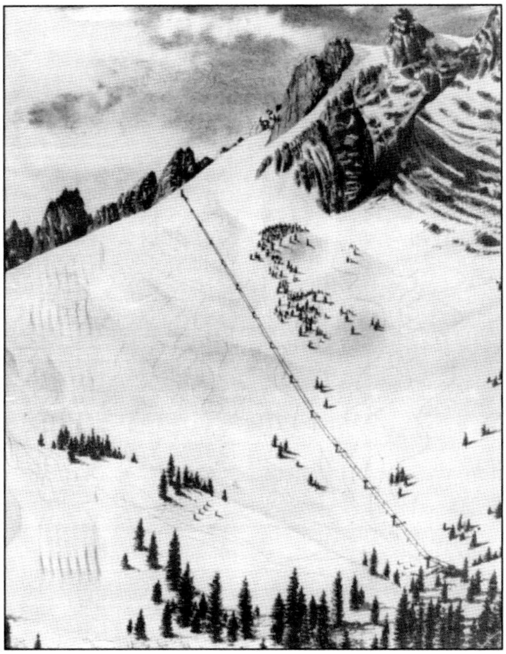

Other entrepreneurs hoped to build ski resorts on the Inyo. One was at Trail Peak south of Lone Pine. The plan included an access chairlift from the Owens Valley and substantial resort condominiums. Forest Supervisor Everett Toll in 1971 followed national policy in not allowing extensive lodging on National Forest land. Another proposal was for Robinson Lake, above Independence. A variety of challenges also stopped this plan. Here is an artist's drawing of one of the main lifts proposed for the slopes of University Peak. The other Eastern Sierra ski area that did catch on was June Mountain, which Bud and Lois Hayward opened in 1961, and is now operated by Mammoth Mountain. (Courtesy of ECM.)

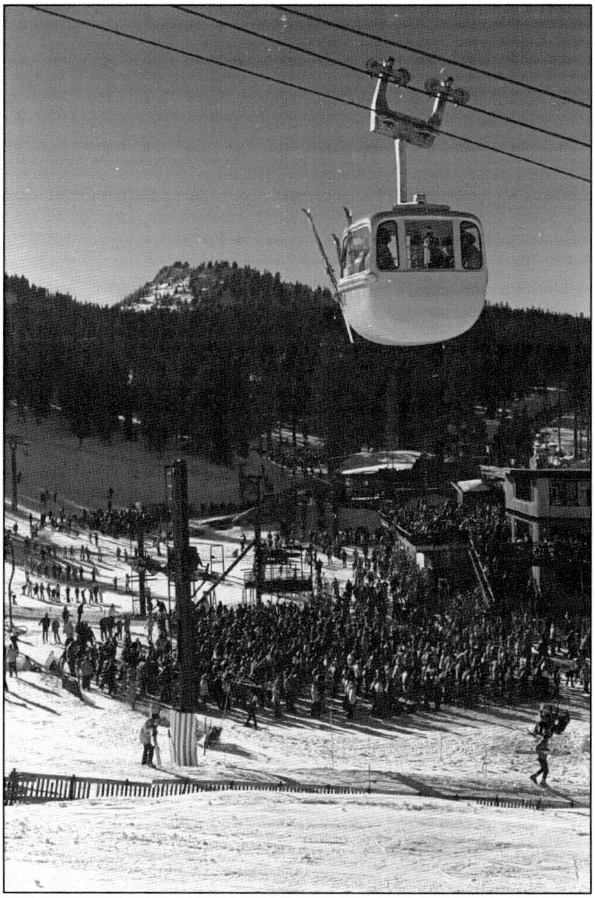

Inyo's forest supervisors supported ski recreation. Rangers, including Rocky Rockwell, Fred Meckel, and Bill Murphy, were avid skiers. Mammoth Mountain's deep snows and extensive slopes, the continuing boom of Southern California, and Dave McCoy's can-do attitude came together to build Mammoth into one of the leading ski resorts in America. By 1967, a gondola ran to the summit. Here, a crowd packs Mammoth's slopes on December 30, 1973. (Courtesy of MMSA.)

At every stage of growth Dave McCoy came up with a way to get the next lift, tower, cable, or structure built and, as over 20,000 weekenders a day flooded Mammoth, he became the most powerful man in the Eastern Sierra. The Inyo National Forest also helped foster the growth of the town by exchanging National Forest land to developers for private inholdings elsewhere. Here, McCoy is pictured at Mammoth in the early 1970s. (Courtesy of MMSA.)

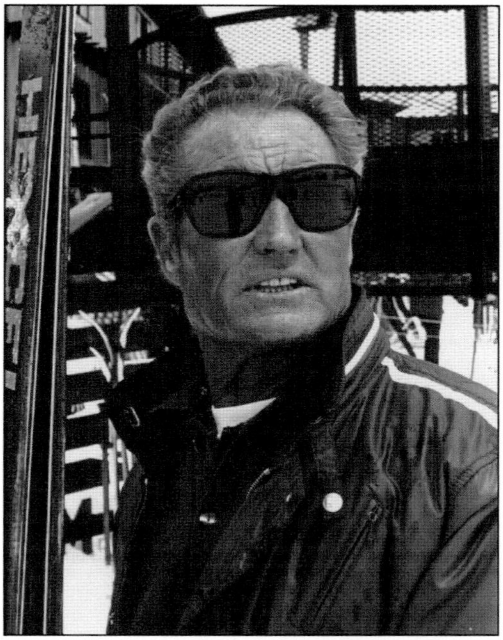

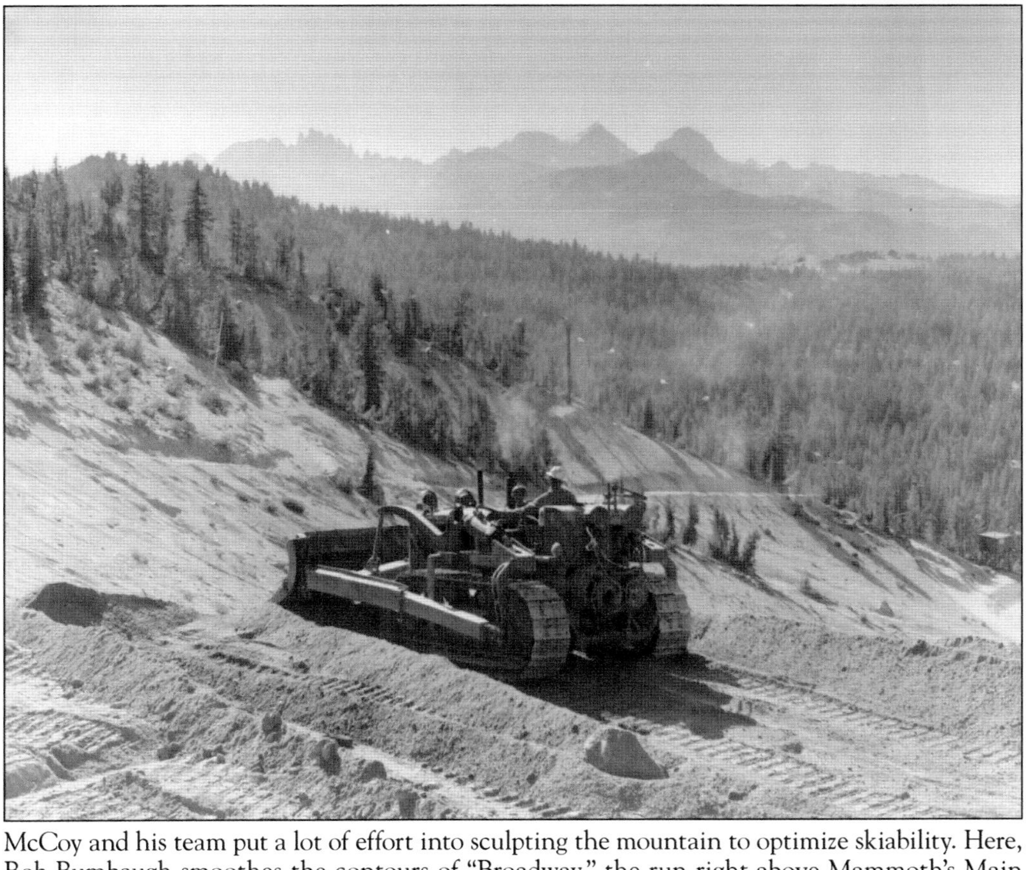

McCoy and his team put a lot of effort into sculpting the mountain to optimize skiability. Here, Bob Bumbaugh smoothes the contours of "Broadway," the run right above Mammoth's Main Lodge. (Courtesy of Dave McCoy collection.)

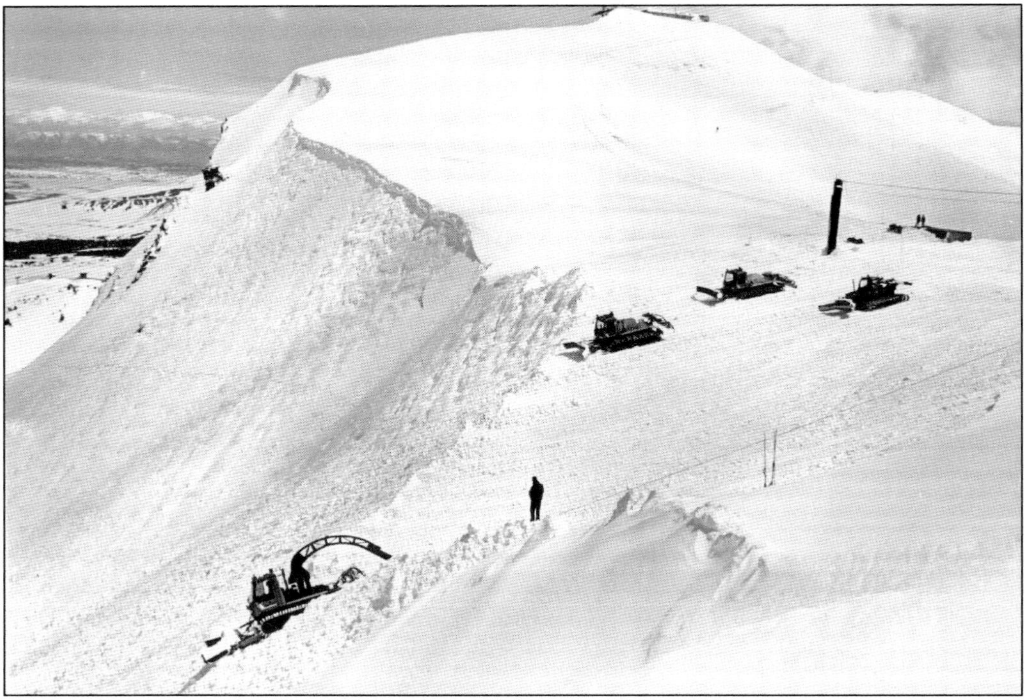

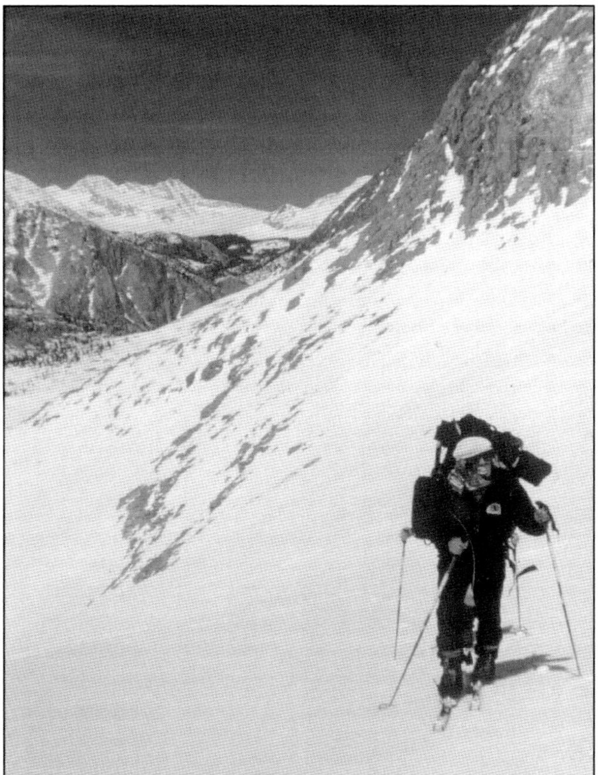

A bold innovation in 1982 allowed CAT drivers to lower off the top and shave down the cornice that formed over "Cornice Bowl." A cable was strung between two anchor poles, and CATs with drivers were then lowered off the edge with a winch to scrape the steep slope, then winched back up. (Courtesy of MMSA.)

In the late 1960s, backcountry enthusiasts started a revival beyond the ski areas. In 1970, Carl "P-Nut" McCoy and Doug Robinson repeated Bartholomew's Whitney-to-Yosemite expedition. A few years later, skier David Beck (left), with Nick Hartsell, reinvented a Sequoia-to-Whitney trip pioneered by Otto Steiner in 1937. Connecting over Shepherd Pass to Independence, their route was dubbed America's "High Route." Beck and his wife, Susan, started a backcountry guide service, Sierra Ski Touring. The upper Kern drainage, shown here behind Mounts Whitney and Tyndall, is some of Beck's favorite ski country. (Courtesy of Chris Cox.)

Open slopes, extensive fair weather, and reliable and relatively safe snowpack make the Sierra one of the best ranges in the world for backcountry skiing. Since the 1970s, "Eastside" skiers have developed gear and techniques to climb, tour, and descend here. Will Crljenko, a one-time Mammoth CAT operator, turned to the backcountry to ski snow, as he put it, "groomed by God." Here he skis from the summit of Mount Dade, above Rock Creek. After thousands of ski days over nearly 40 years, Crljenko and a partner were killed in 2005 in his first avalanche incident. (Courtesy of AS.)

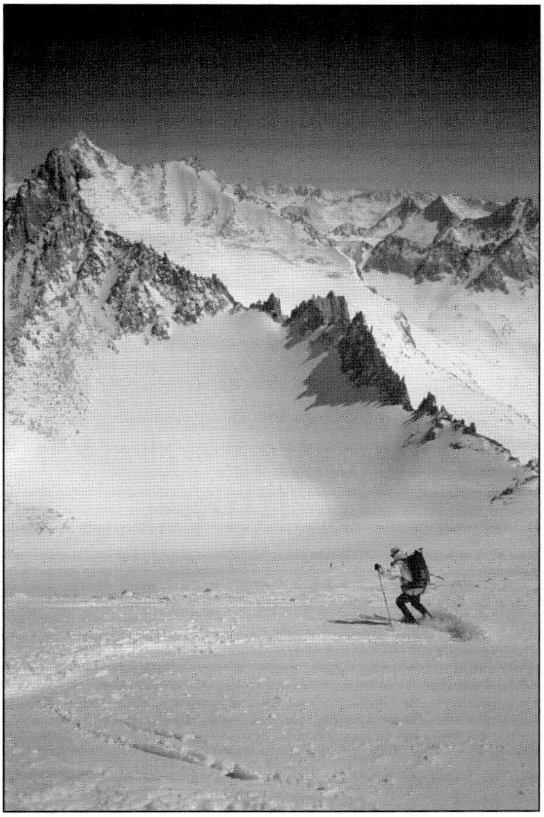

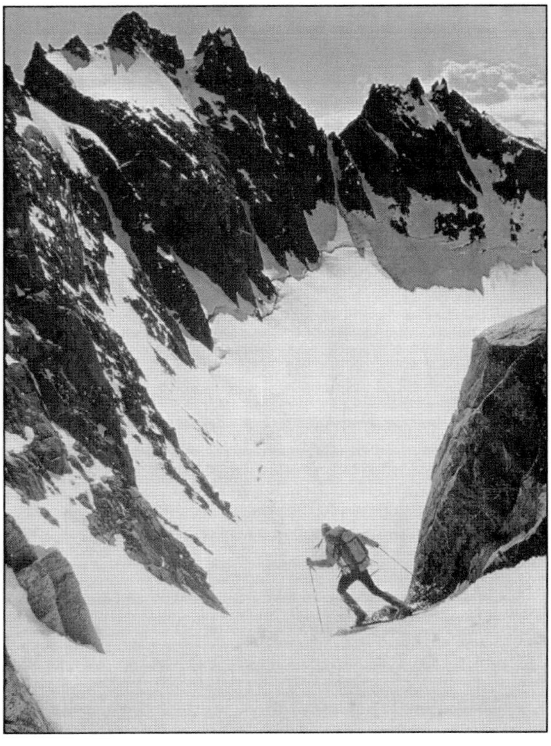

In the 1970s and 1980s, backcountry skiers focused on becoming proficient with lighter weight gear to cover more extensive terrain and still ski "the steeps." One of the most impressive "skinny ski" routes was the "Redline" tour pioneered by Tom Carter, Allan Bard, and Chris Cox, joined by a couple of others at times. Over two great snow seasons, 1982 and 1983, they followed the Sierra Crest as closely as possible, over the peaks, from Mount Langley north to Mammoth. Here, Chris Cox swings turns onto the Palisade Glacier from Mount Sill. (Courtesy of Chris Cox.)

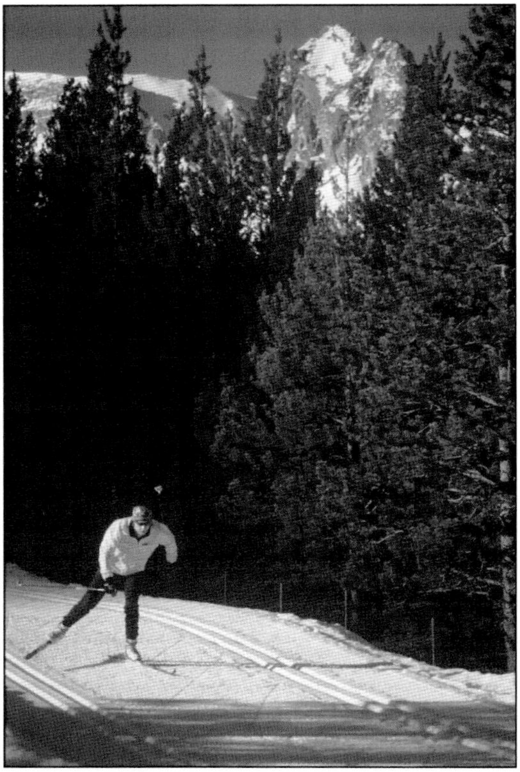

In the 1970s, Tamarack Lodge owner Bob Stanford began keeping the lodge open through the winter and started a Nordic ski operation. Today, this is one of California's favorite cross-country ski centers, with groomed trails for touring and racing and a growing biathlon program. Local resident and two-time Olympic racer Nancy Fiddler has inspired and taught a whole generation of Nordic skiers here, focusing on both "masters" training and kids' programs. At left, Nancy speeds down the Lake Mary Road on a workout. Below, the "gliders" program she started gets kids from Bishop to Lee Vining out in the snow and culminates in a March fun-race. (Courtesy of AS.)

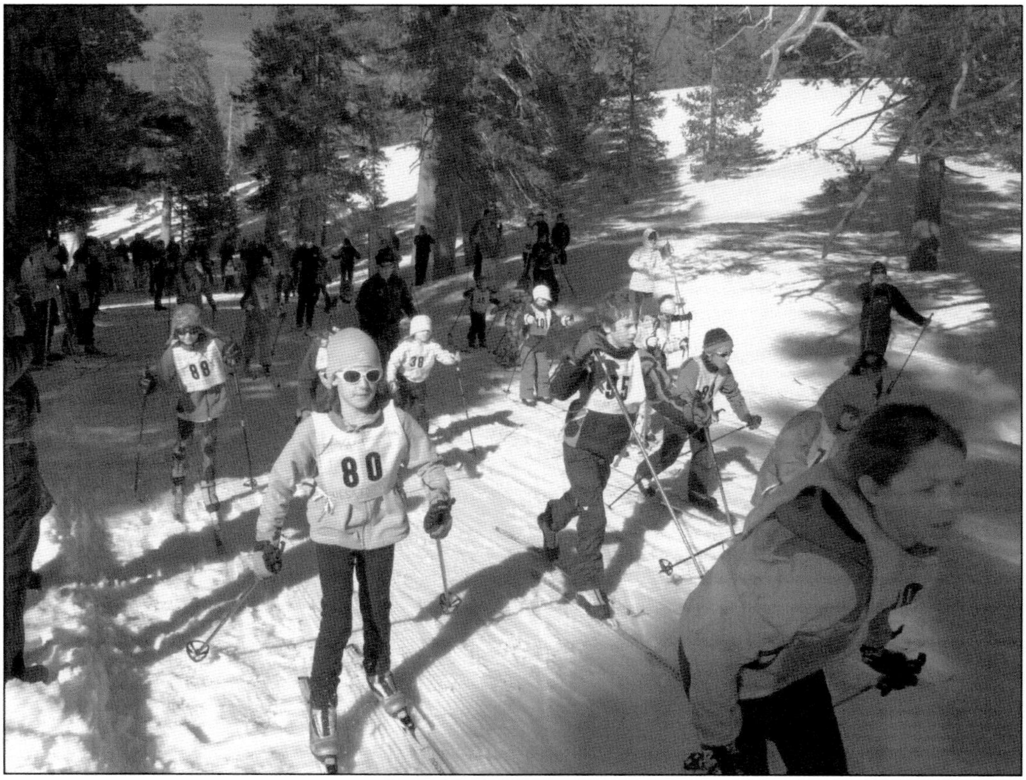

Seven

Taking on the Up

Climbing

Ever since surveyors in the 1850s found that the Southern Sierra held the highest mountains in the United States, Sierra peaks have been a magnet for people to lift their bodies and spirits to as high as the land will take them. With row upon row of peaks scraping a usually friendly sky and routes from easy to challenging on generally solid rock, the High Sierra developed a reputation for some of the most rewarding and scenic climbing in America.

The Sierra's friendly character has encouraged climbers to venture onto the peaks' steeper facets. Particularly in the 1930s, Sierra Club members from both Southern and Northern California began to pioneer ascents up exposed and intimidating buttresses and faces. They enhanced their confidence by developing safety backup techniques on smaller, hometown crags. This lineage that began in the Sierra migrated to the sheer walls of Yosemite and, eventually, to the world's "Great Ranges." Starting in the late 1950s, Yosemite veterans metaphorically returned to the Sierra peaks to climb the sheerest, Yosemite-like walls of the high country. Also at this time, the Angeles Chapter of the Sierra Club solidified its yen for "peakbagging," establishing in 1955 the Sierra Peaks Section of its chapter and a list of 248 worthwhile Sierra peaks to climb.

This legacy of Sierra climbing is dominated by a singularly dedicated man, Norman Clyde. He was a scholar of classic literature who turned away from academia in 1910 and took his first Sierra climbing tour, from Yosemite to Whitney. In the late 1920s, he turned away from nearly all social ties, and made the Sierra his hermitage. Clyde climbed peaks into the late 1960s and studied every aspect of the range. Arguably, he came to know these mountains better than anyone in history.

Today, people of all ages, talents, and styles climb in the Sierra. From adventurous hikers on the Mount Whitney trail to athletic wizards who take on the steep faces and buttresses and backcountry enthusiasts who simply find fulfillment in going up different peaks each summer, thousands of people know Sierra climbing as some of the best times of their lives.

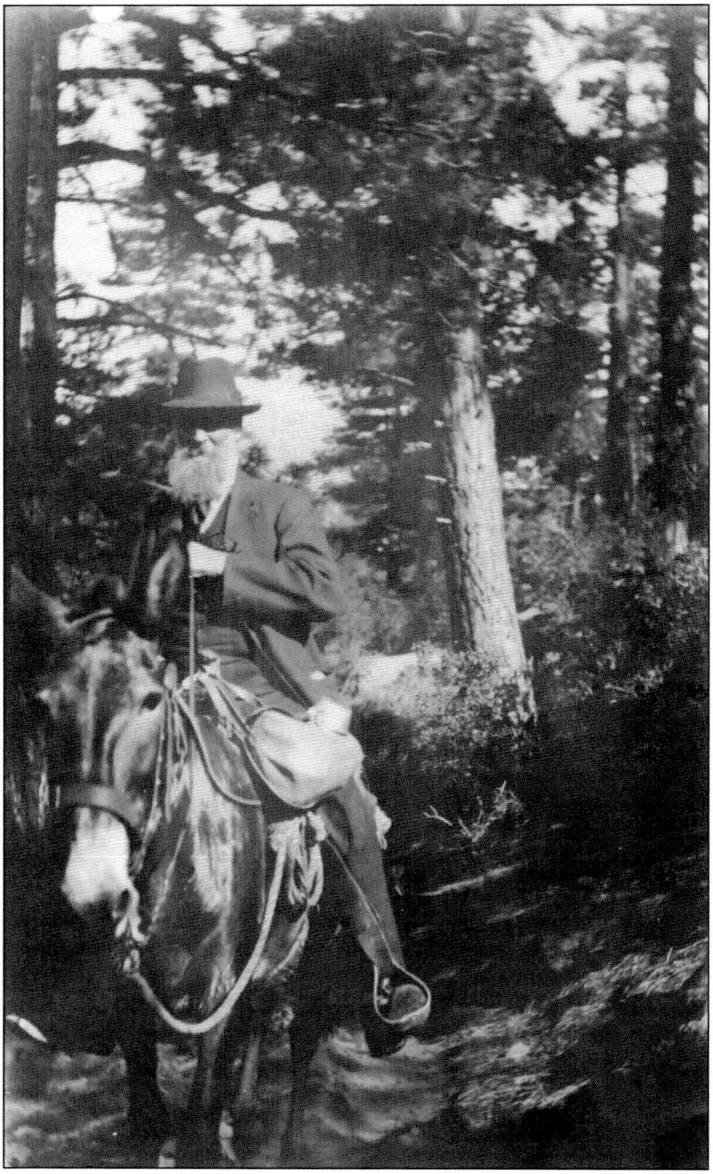

John Muir rides a mule along the Little Kern River around 1908 at about age 70. It was his penchant to climb Mount Whitney that first brought him down the arid east side of the Sierra in October 1873. He had heard how a group of fishermen from Lone Pine had recently made Whitney's first ascent via a roundabout tour over Cottonwood Pass to the southwest slopes. Muir tried to make a fast trip that way but ended up overextended, benighted, and dancing to keep warm at 14,000 feet. He recouped in Independence for a day, then "set out afoot for the summit by direct course up the east side." His route up a chute and onto Whitney's north face must have looked incredible to most people then. He later wrote, "Well-seasoned limbs will enjoy the climb of 9,000 feet required for this direct route, but soft, succulent people should go the mule way." Muir is remembered most, of course, for sharing his utterly inspired joy for roaming immersed in the High Sierra: "Who wouldn't be a mountaineer! Up here all the world's prizes seem nothing." (Edward Hughes photograph, MSS048.f24-1320, courtesy of John Muir Papers, Holt-Atherton Special Collections, University of the Pacific Library, © 1984 Muir-Hanna Trust.)

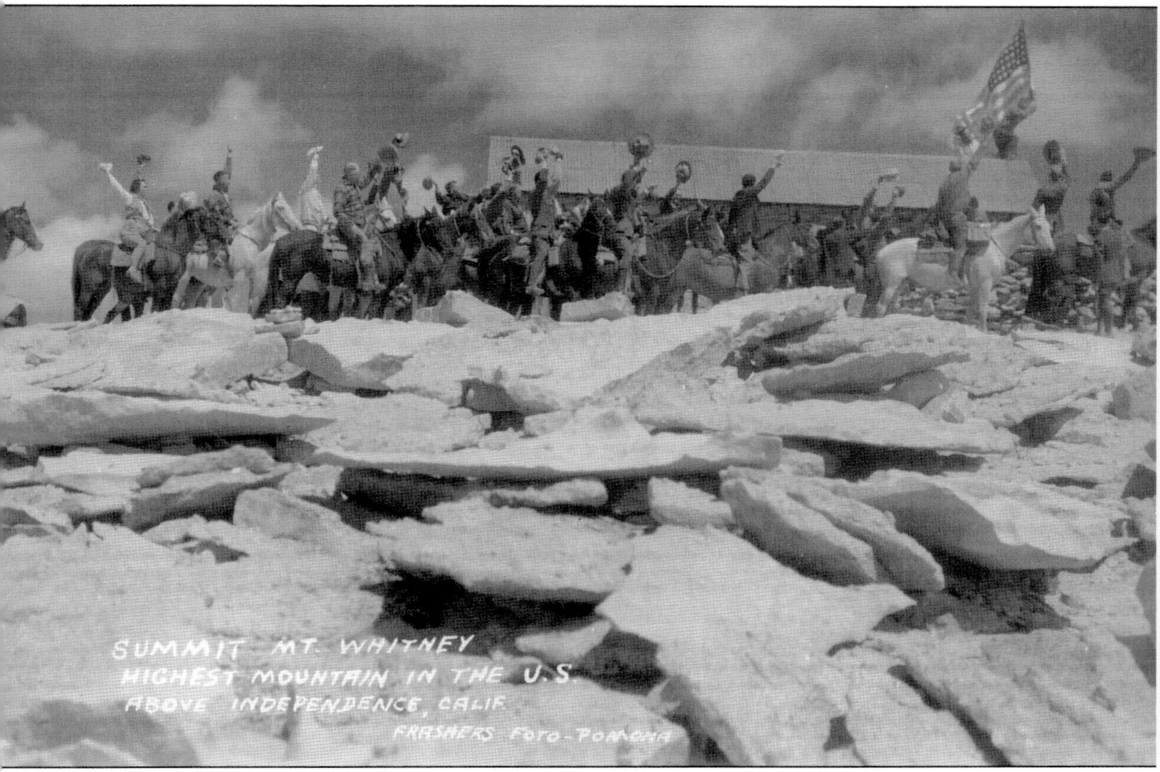

A party on horseback hails the raising of the American flag on Mount Whitney's summit hut in the 1920s. Since 1904, a "muleway" of switchbacks and ledges blasted from granite allowed hikers and stock to climb through intimidating terrain direct from Lone Pine. The hut was built in 1909 for storage and shelter to support the Smithsonian Institution's studies of Mars. (Lightning has killed climbers inside since.) Gradually, the path to the highest point in the United States became a pilgrimage for hundreds every summer. As the Inyo took responsibility for maintaining wilderness values, the heavy use needed to be regulated. Today, the agency issues Whitney trail permits by lottery and limits use to 60 overnight climbers and 100 day-climbers per day, with no mules. (Courtesy of ECM.)

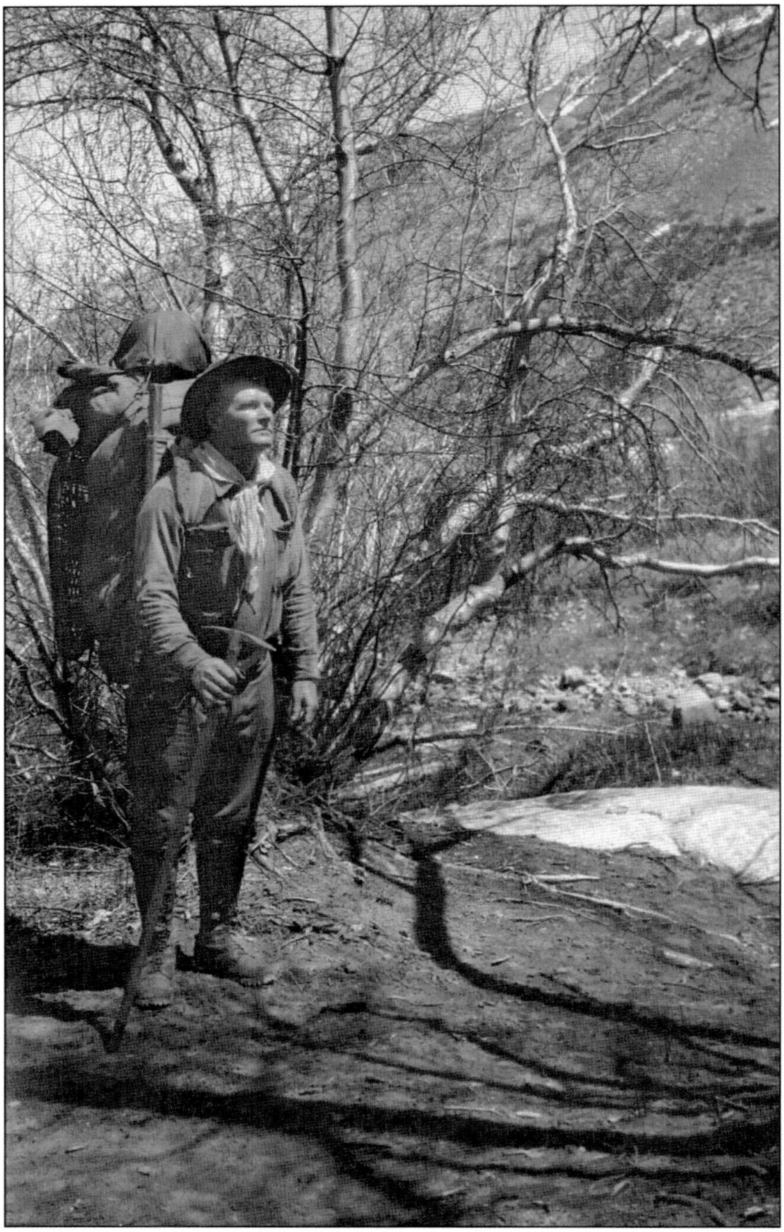

Like many, Norman Asa Clyde found his muse in the Sierra and, unlike anyone else, he held fast to his passion and lived in these mountains. And, he climbed practically all of them. He was an avid climber and schoolteacher around the West and, in 1926, he landed a job in Independence. A year later, his temper and fondness for guns led to warning shots fired at rowdy students, which ended his teaching career. He took to living out of his enormous backpack in the Sierra for half of the year and caretaking at trailhead lodges in the winter, particularly at Glacier Lodge. He made the first ascents of about 117 Sierra peaks. Heading into the mountains for weeks at a time, his packs were essentially households and libraries, including cameras, pistols, an anvil for boot repair, and cast-iron pans; as a scholar in classic literature, he carried thick volumes in Greek, Latin, and other languages. Here, he hikes into the Palisades from Glacier Lodge. (Courtesy of Norman Clyde collection.)

Norman Clyde was paid a stipend to lead climbs on many Sierra Club High Trips. He would escort trains of novice climbers up easy routes and would go with the club's most skilled on wildly exposed routes, usually first ascents. Here, he sends a rope down for others from the top of the infamous "Milk Bottle" summit block of Starlight Peak, a 14,003-foot-high peak in the Palisade chain. Below him, Sierra Club president Francis Farquhar claws his own way up. (Courtesy of Glen Dawson collection, Angeles Chapter of the Sierra Club.)

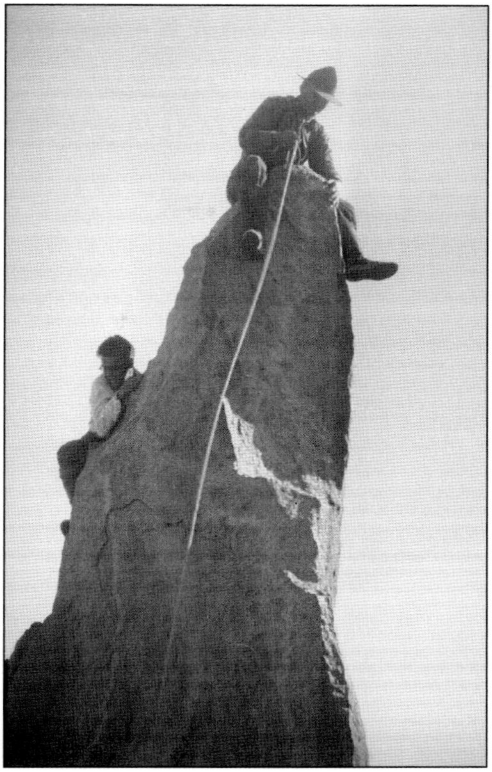

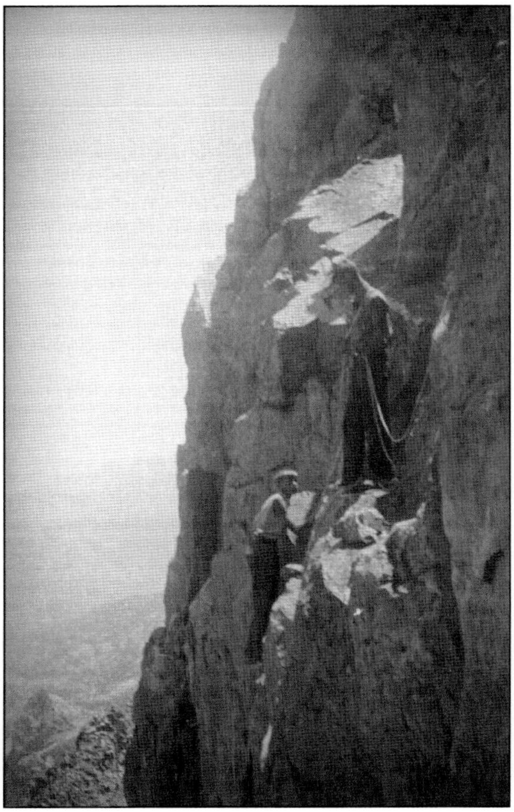

In the 1930s, a clan of young Sierra Club climbers took to the peaks with aplomb, especially after they learned rope techniques from Norman Clyde and Harvard professor Robert Underhill. Here, Jules Eichorn belays Walter Brem on Michael Minaret, a peak in the Ritter Range out of Mammoth. Unfortunately, this 1933 trip was a search for a missing comrade, Walter Starr Jr., who had gone on a solo-climbing trip. Norman Clyde later found Starr's fallen body not far from where this picture was taken. (Glen Dawson photograph, courtesy of Angeles Chapter Sierra Club.)

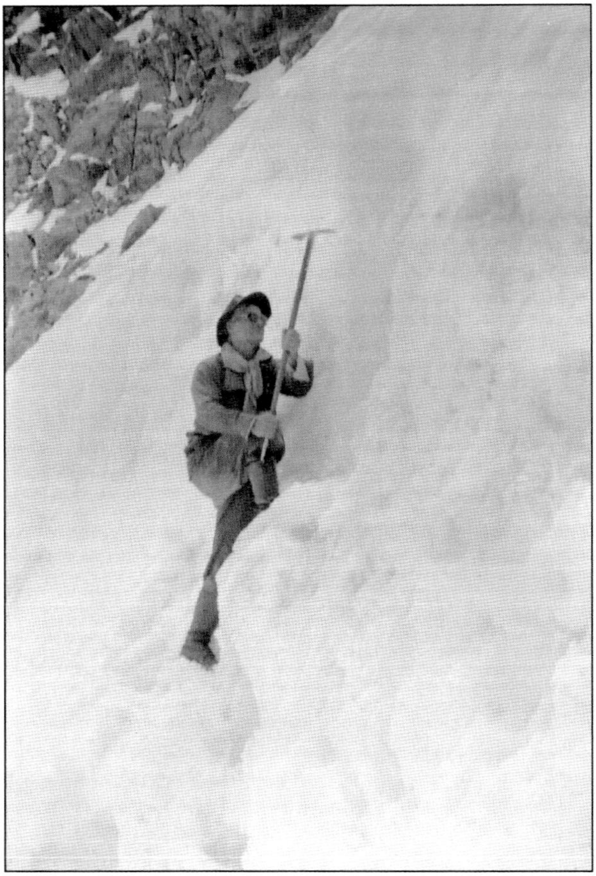

Having learned from Swiss guides in Canada, Norman Clyde was unique among Californians in his ability to navigate on snow and ice as well as rock. He is pictured on a route he pioneered, the U-notch couloir on 14,242-foot-high North Palisade. Clyde was a connoisseur of mountains, not a seeker of notoriety; climbing was his way to participate with the earth's most exalted terrain. (Courtesy of Norman Clyde collection.)

Another early settler in the Owens Valley was William "Smoke" Blanchard, who came to Bishop in 1942. A Buddhist, bard, truck driver, and charismatic adventurer, Blanchard looked at mountains with an original eye. He climbed mountains from Alaska to Japan and enjoyed exploring the crags and boulders of "Buttermilk Country" above Bishop. He also became one of Clyde's few close friends. Here, he stands on the "Peabody" rocks of Buttermilk Country with his wife, Su, and children, Glen and Lori. (Courtesy of Ron and Nancy Smith.)

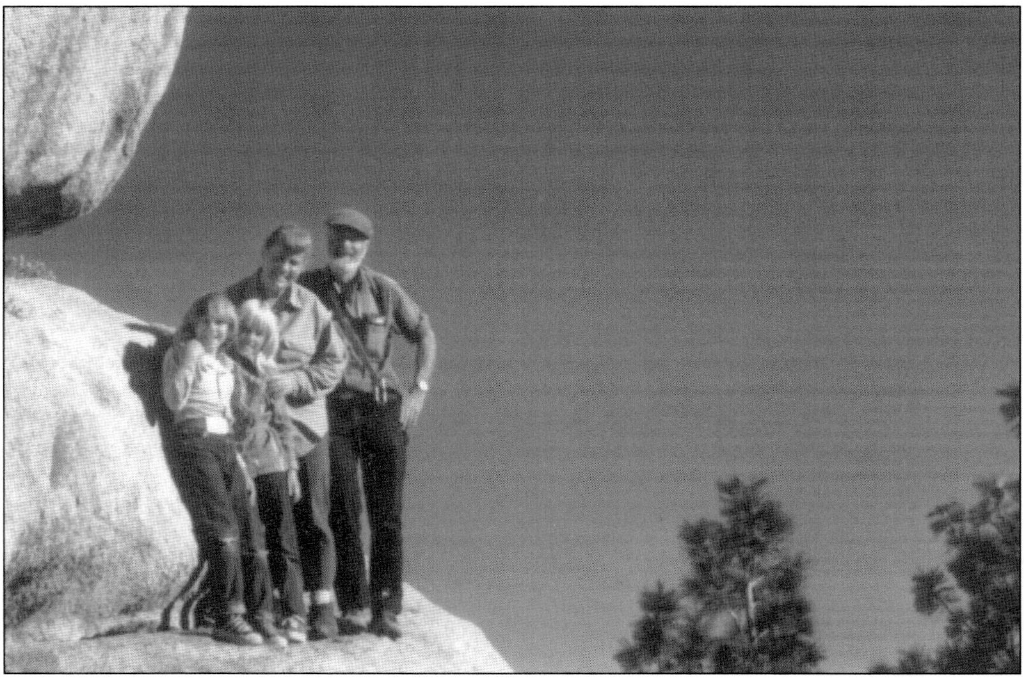

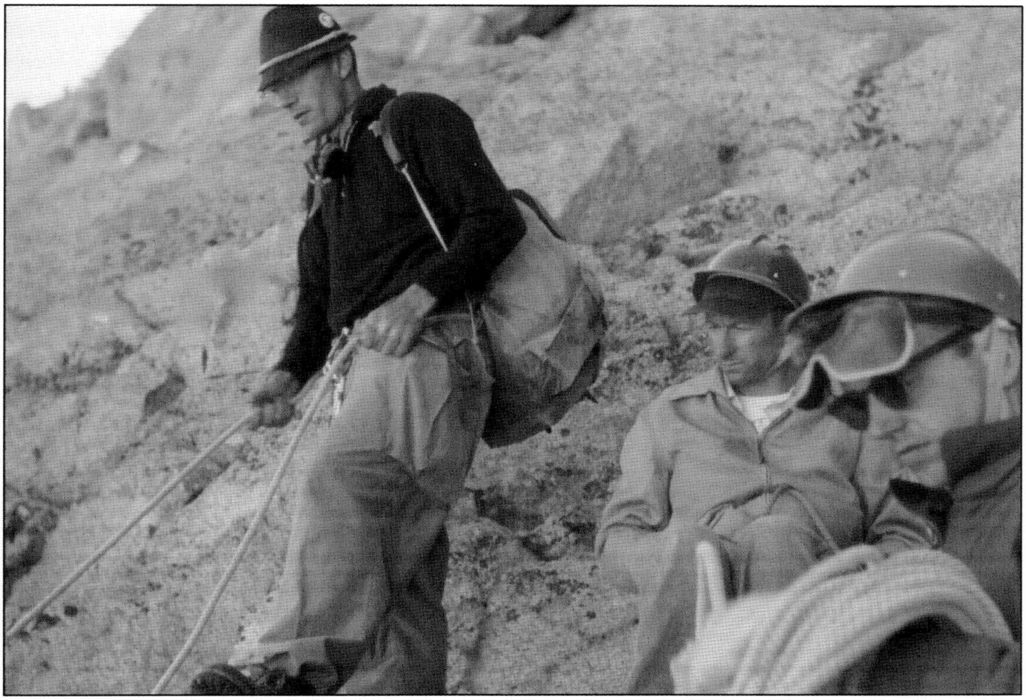

Larry and Laurie Williams were an adventurous couple with lots of Sierra under their belts and, in 1959, they started Mountaineering Guide Service (MGS), America's first official climbing school and guide service on National Forest land. Larry led aspiring climbers up various peaks, especially in the Palisades. Here, he belays clients in 1961. (Courtesy of Laurie Williams.)

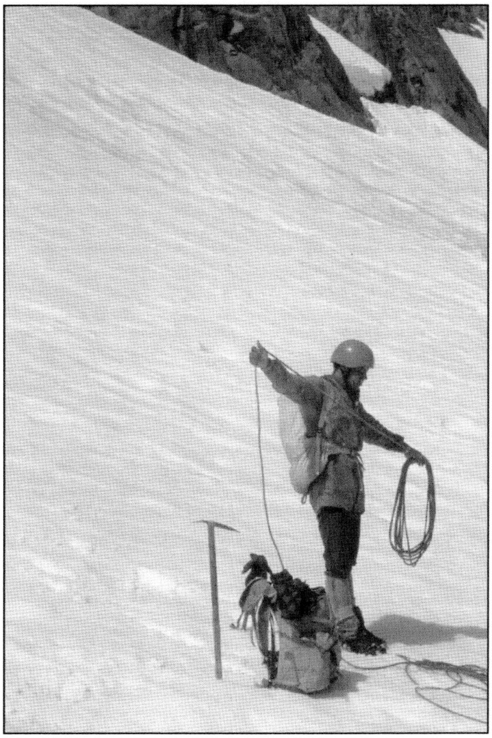

The most prominent guide hired by the Williams was Don Jensen, a young man from the Bay Area. Jensen had dropped out of Harvard to base the whole winter of 1962–1963 out of a tent among the Palisades. He spent several seasons on avant-garde climbing expeditions in Alaska, working the second half of his summers as a guide for MGS. He also developed many new routes up buttresses in the Palisades and innovations in packs, tents, and ice screws. Sadly, both Larry Williams and Jensen died young. Williams died when both engines on his Cessna failed at takeoff in 1967, and Jensen passed away due to a bike accident while at university in Scotland in 1973. (Courtesy of Keith Brueckner.)

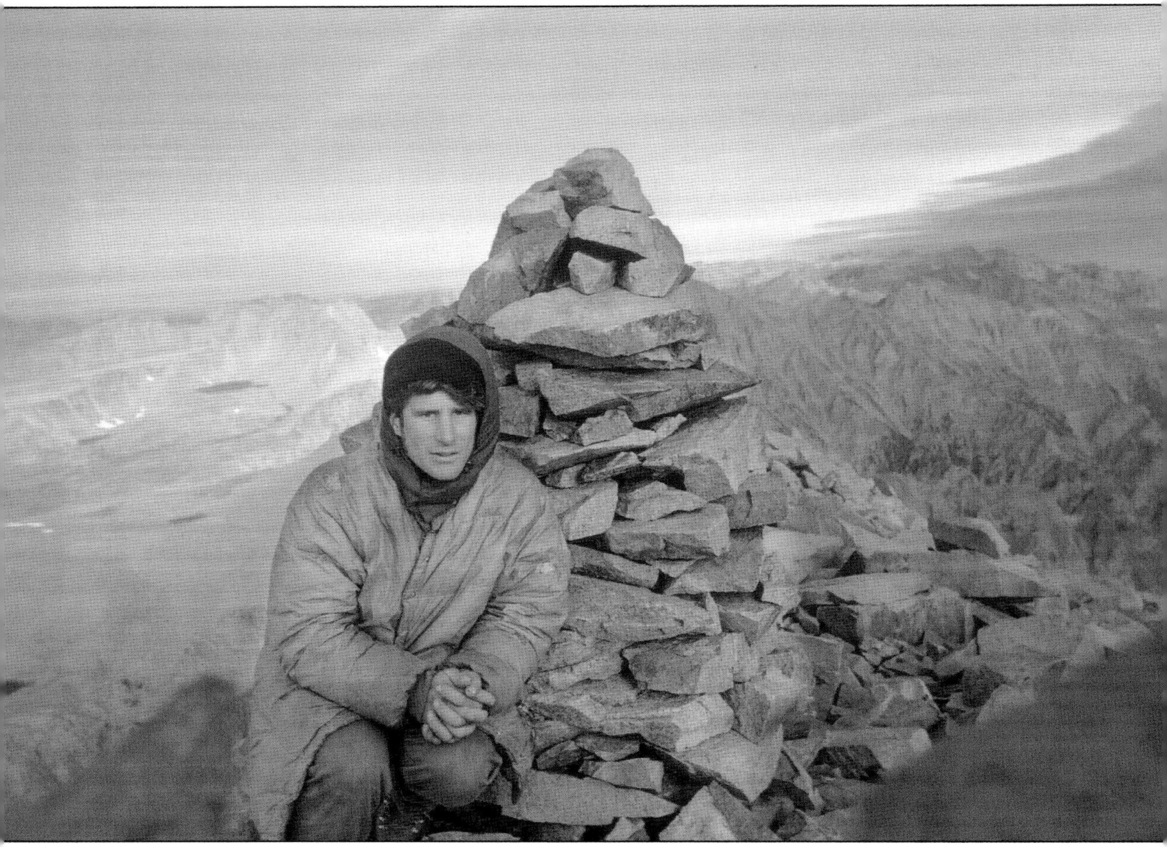

Galen Rowell began a lifelong career of mountain adventures, photography, and writing in the Sierra. From his Berkeley home, he blasted over Tioga Pass for countless endurance weekends. He climbed with Warren Harding of El Capitan fame and took up Harding's torch for bringing Yosemite skills to High Sierra walls. Here, Rowell is alone on top of 14,375-foot Mount Williamson in the fall of 1974, a few months before heading to K2—his first of many Asian climbing expeditions. His many adventures on the Inyo include new routes up Mount Whitney and 14,064-foot-high Split Mountain and the first ski tour along the top of the White Mountains. (Courtesy of Galen Rowell/Mountain Light.)

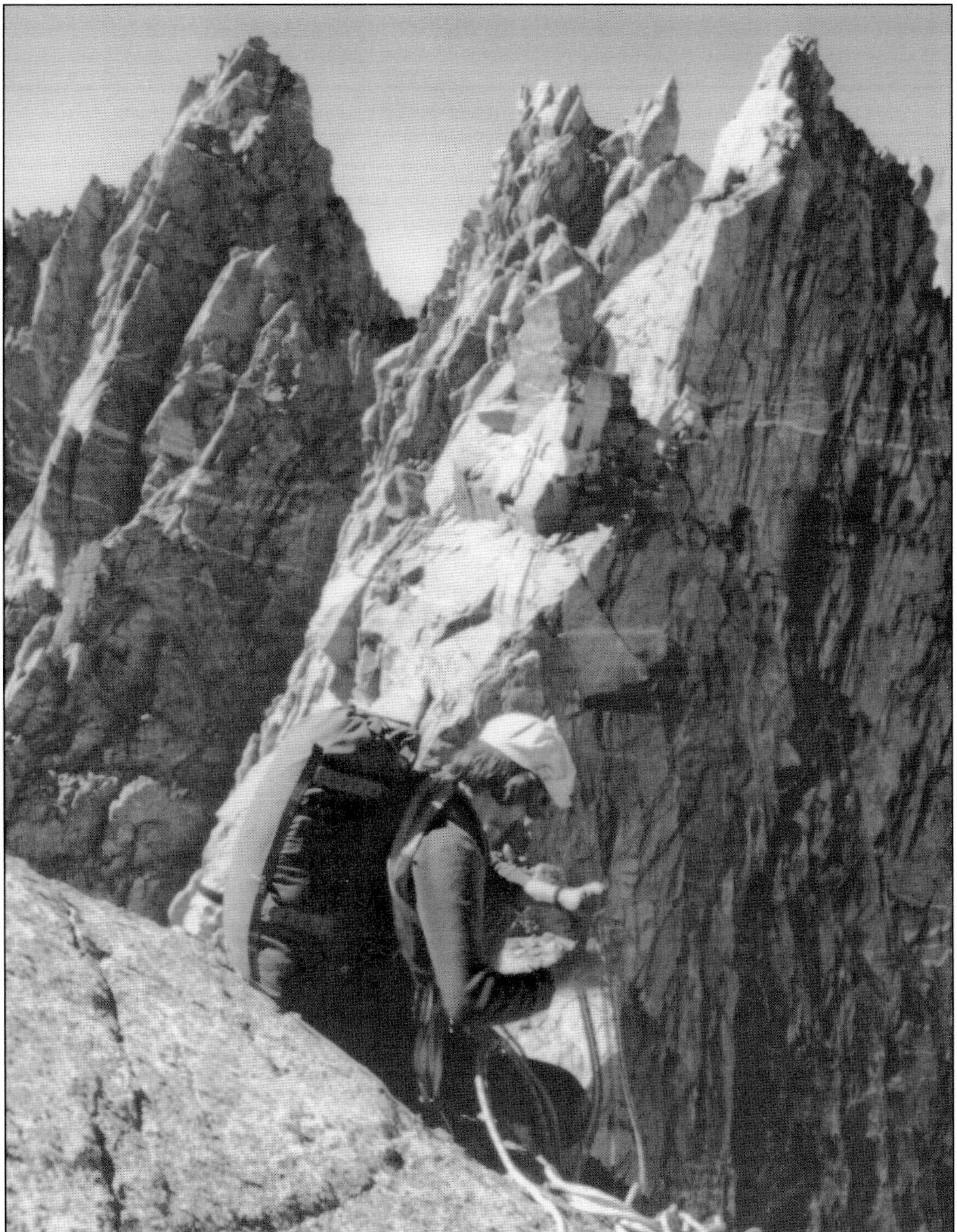

Many Sierra peaks are part of a saw-toothed chain, and climbers since the 1930s had seen the beauty in extended scrambling in the sky from peak to peak. Don Jensen had mused about traversing the entire Palisade chain, from Southfork Pass, over some eight miles of picket-fence peaks over 13,000 feet high, to Bishop Pass. In 1979, guide John Fischer and one of his clients, Jerry Adams, completed that. Adams set 3 caches along the foot of the peaks, and over 12 days, the two pulled it off. Here, Fischer arranges a rappel about a third of the way through the chain. (Courtesy of Jerry Adams.)

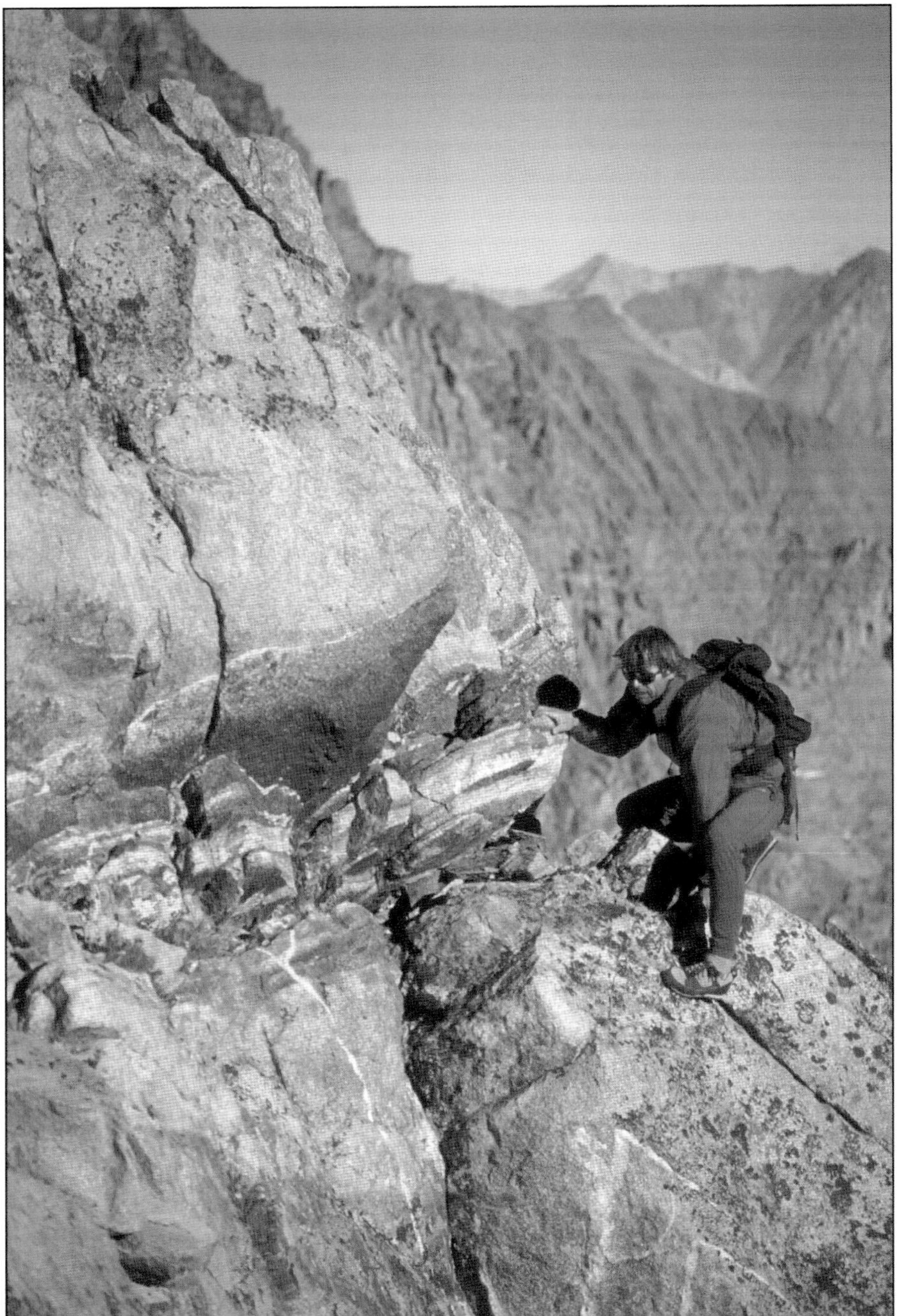

Other talented climbers brought their Yosemite-honed skills to Sierra traverses. Notably, in 1982, Claude Fiddler and Vern Clevenger in two long days climbed over the top of all of the Minarets. In 1992, Canadian-born Peter Croft repeated that traverse in an epic day from the car. Croft was widely regarded as the best rock climber in the world, but he turned away from the trend of seeking sheer gymnastic difficulty in favor of extending the fun of climbing peak after peak on what is for him moderate terrain. Here, Croft traverses peaks in the southern group of the Palisades. (Courtesy of AS.)

Eight

The Mountains Resonate

Art

A powerful landscape like the Eastern Sierra calls people to explore not only the land but also the terrain of their minds, particularly responses to beauty and their position in the natural world. While of course the Paiutes had spiritual and artistic roots here for centuries, early Euro-Americans out West hoped for landscapes that were verdant, pastoral, or well glaciated, like in Europe. The large, hard-edged scenes of the Inyo at first seem vacant and intimidating. But artists, as much as hikers, have found that this land is a magnificent and welcoming realm when it is met on its own terms.

Writer Mary Austin was an early artist to fall in love with American deserts. She settled in Independence and helped initiate a fascination for the native peoples and landscapes with her 1903 book, *Land of Little Rain*. The Owens River diversion left her frustrated and heartbroken, and she took her passions to Carmel. There and in New Mexico she met many artists, including photographer Ansel Adams.

A variety of notable artists found inspiration in the Inyo during the early decades of motor tourism, including Burton Frasher, Stephen Willard, and Robert Clunie. In the 1940s, Ansel Adams created some of his most iconic landscape photographs here. More than any other artist of the era, he transmitted the Sierra into the American mind and inspired an optimism that a new and grander world was waiting in the bold contours of the West. A generation later, Galen Rowell added a keen photographic eye to his athletic adventures and, with the Eastern Sierra his home range, he inspired a new generation of mountain photography.

Hollywood also discovered the Sierra. Right from the silent movie–era, directors arranged Western and Far North sets here. Also, companies ranging from Marlboro to BMW focused advertising campaigns using the scenery of the Inyo. Musicians are also readily inspired to write and perform here. Today, the Eastern Sierra is one of America's favorite destinations for photography workshops and all kinds of terrain-based artistry.

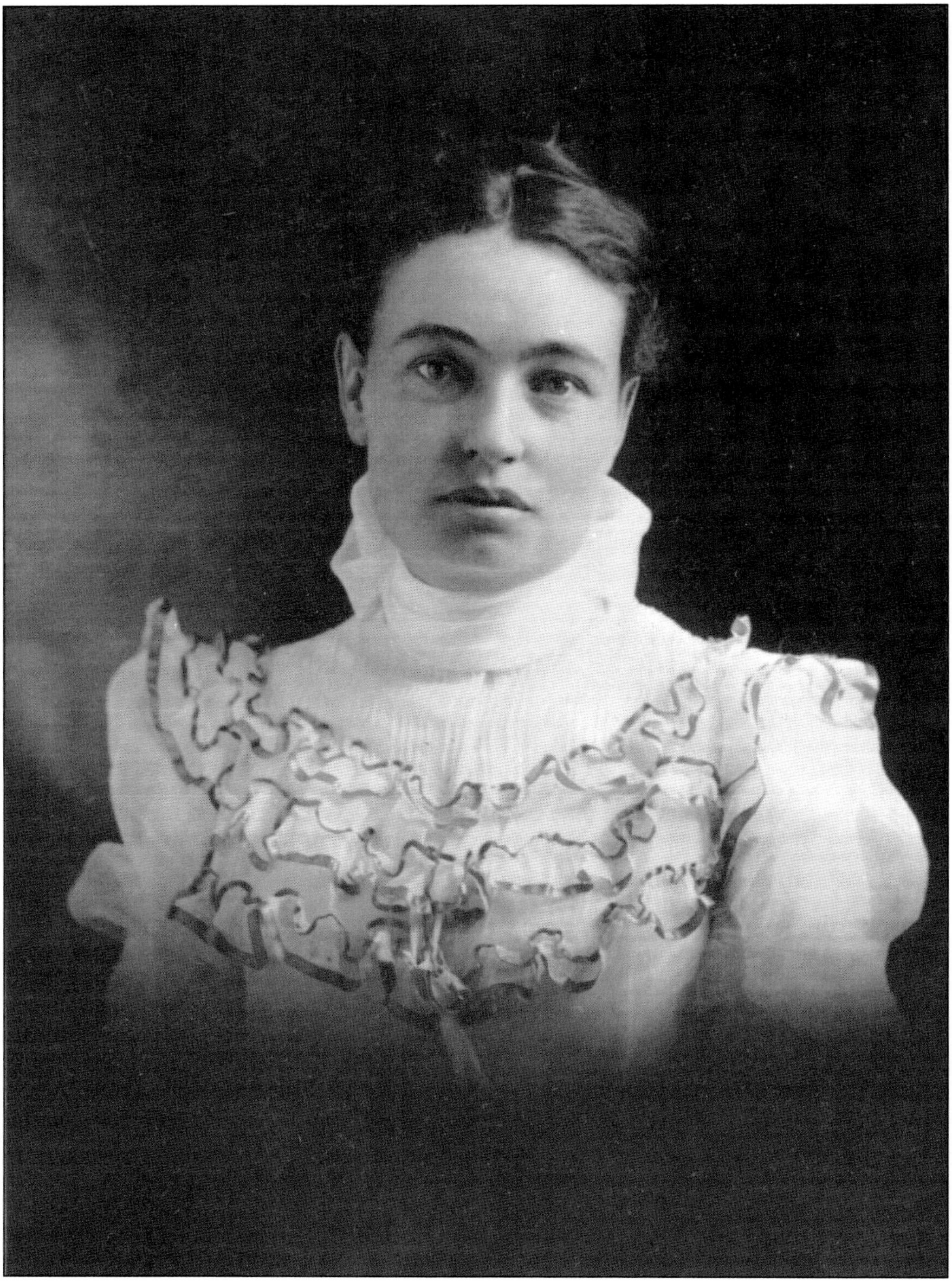

This 1903 portrait of Mary Austin is from the year she published *Land of Little Rain*. The mountains above her home in Independence inspired her to write, "Let's go back across Mohave where the hills of Inyo rise, there's a word we've lost between us we shall never hear again, in the mindless clang of engines where they bray the hearts of men. Let's go seek it east of Kearsarge where the seven-mile shadows run, from the great gray bulk of Williamson heaved up against the sun." (Courtesy of ECM.)

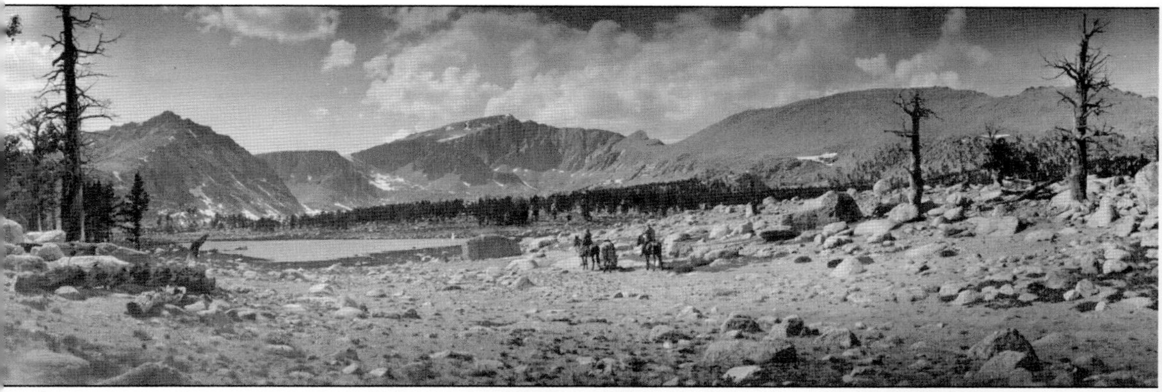

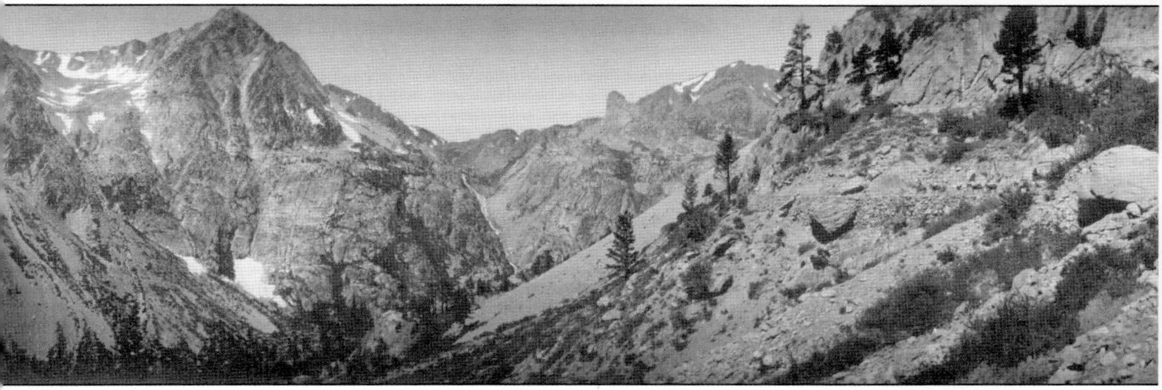

Matt Ashby Wolfskill was a little-known German-American photographer who lived in Southern California. In 1911 and 1912, he came up to the Sierra in a horse-drawn wagon with photographic equipment, notably a large panorama camera. The first image at the Cottonwood Lakes, showing 14,025-foot-high Mount Langley. The second image shows his pack train heading up the road up Lee Vining Canyon to Tioga Pass. The panoramic format emphasizes Wolfskill's apparent love for a classic experience of the West—engulfed in vast grandeur. Across Tioga Pass he would soon make one of his most important photographs, of Hetch Hetchy Valley a few years before it was dammed. (Courtesy of ECM.)

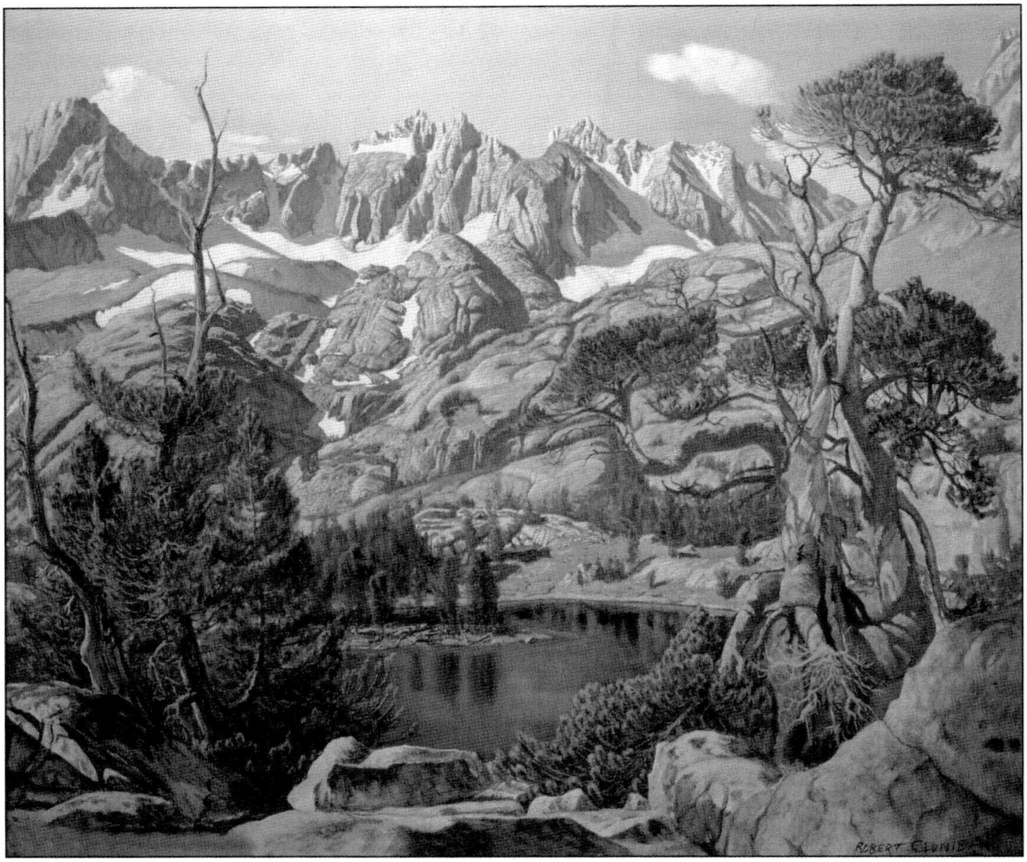

Robert Clunie became the most well-known painter of Eastern Sierra landscapes. Born in Scotland in 1895, his interests drew him progressively west to find his soul-place beneath the Palisades above Big Pine. As he built his fine art career, he also worked in Hollywood and Santa Paula painting movie sets, houses, train cars, and the like. In 1928, he and his wife ventured up to the Sierra and they returned to Yosemite and Mammoth the following year. When they walked into the Big Pine Lakes, he decided to stay for eight weeks and completed at least 14 renditions of the scenery there. He would return to his campsite between Fourth and Fifth Lakes some 30 times, creating dozens of oil paintings there, as always, *en plein air*. Here is his work, *Fourth Lake—North Palisade Region* (1950). Below, he begins another work at Fifth Lake, with Temple Crag and Mounts Gayley and Sill behind. (© 2011 Coons Gallery/Spotted Dog Press Inc.)

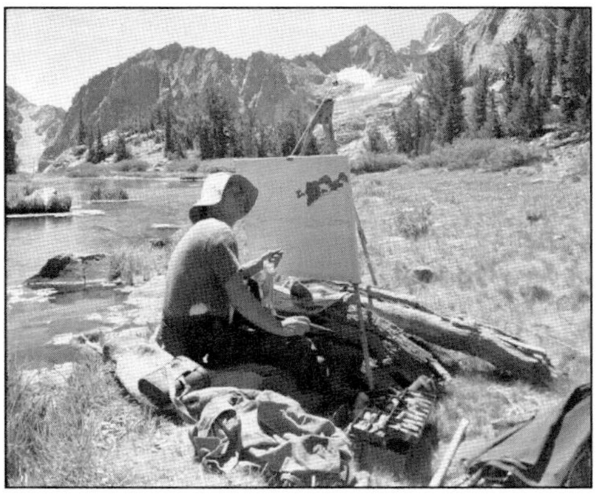

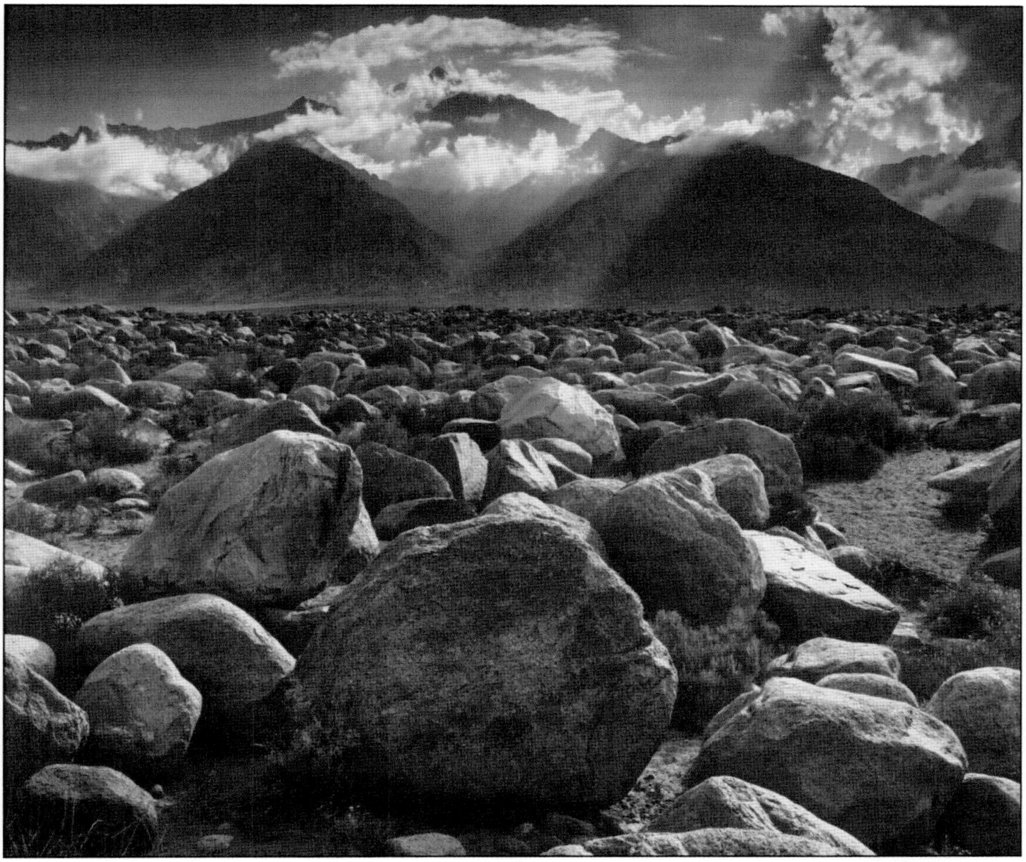

During World War II, Ansel Adams, under contract with US Camera and the War Relocation Authority, photographed life at the Manzanar War Relocation Center. Outside the camp, he set his camera toward the sun descending through the last clouds of a storm over Mount Williamson. He soon wrote, "In these years of strain and sorrow, the grandeur, beauty, and quietness of the mountains are more important to us than ever before. I have tried to record the influence of the tremendous landscape of Inyo on the life and spirit of thousands of people living by force of circumstance in the Relocation Center of Manzanar." This photograph became one of the most iconic masterpieces in his career. With severe, backlit contrast and the dynamic clouds constantly shifting the light, it was a technically challenging scene to capture well. Yet it happened, because the landscape and the craft and vision of the artist came together. (Photograph by Ansel Adams, courtesy of Collection Center for Creative Photography, University of Arizona, © The Ansel Adams Publishing Rights Trust.)

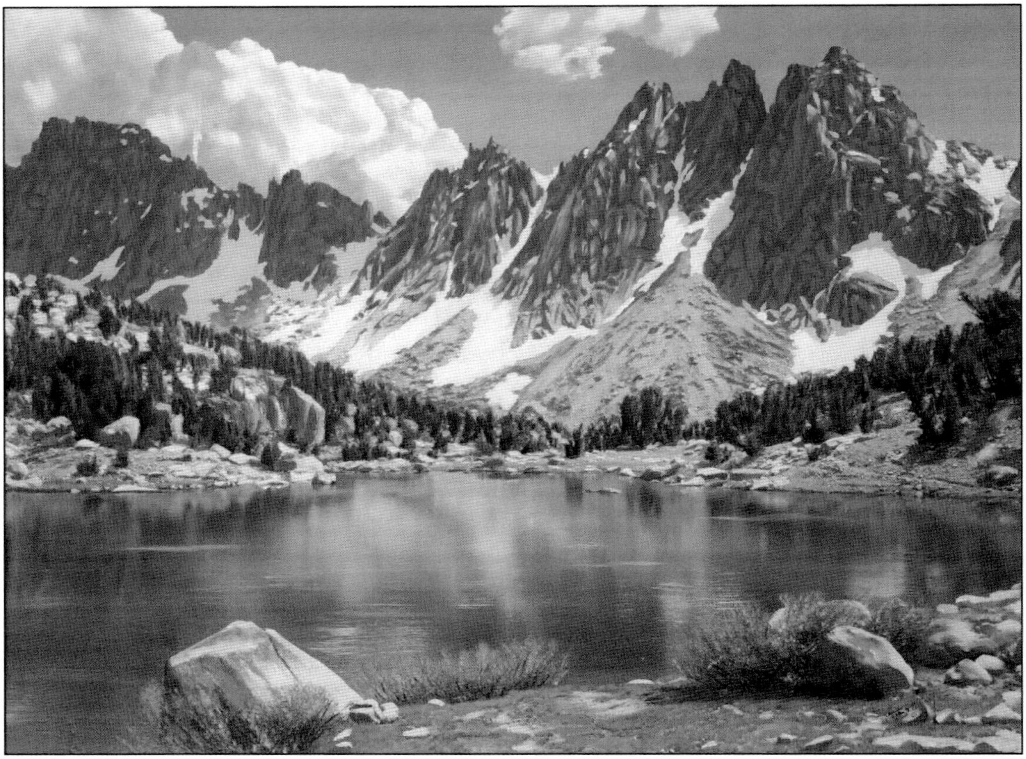

Stephen Willard (1894–1965) was a renowned landscape photographer-painter who built an extensive body of work in the California deserts and the Sierra. From at least the early 1930s, he started hand-tinting his photographs to excellent effect. Over decades in his work, one sees a progression from subtle washes to fairly dense coloration. In 1934, the Inyo gave Willard a permit to build a recreational residence near Twin Lakes above Mammoth, and in the summers he both lived and sold his work there (below). The current owners of the cabin still run it as a seasonal gallery, probably the only such operation on National Forest land. Above is one of his most favored Sierra works, of the Kearsarge Pinnacles above Independence. (Above, courtesy of Jeff and Maranda Moran collection; below, courtesy of AS.)

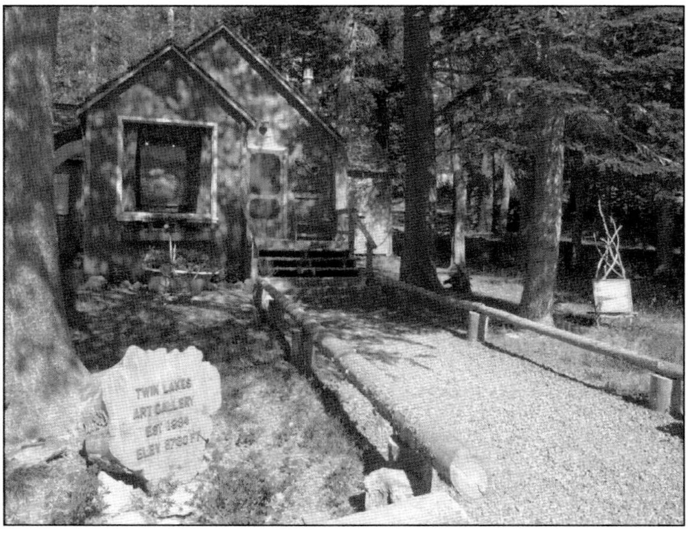

This publicity photograph from the 1928 silent movie *Back to God's Country* was taken in the meadows below the Sherwin Ridge, near Mammoth Creek. Actors Renee Adore and Robert Frazer are set to travel across the arctic by dogsled. The dogs and sled were rented from Tex Cushion, who acquired them from an earlier film team that left them behind. (Courtesy of Beverly and Jim Rogers Museum of Lone Pine Film History.)

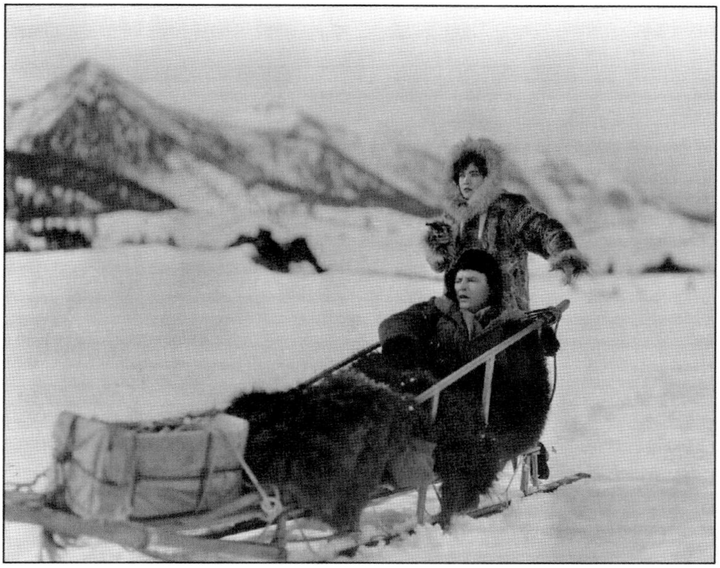

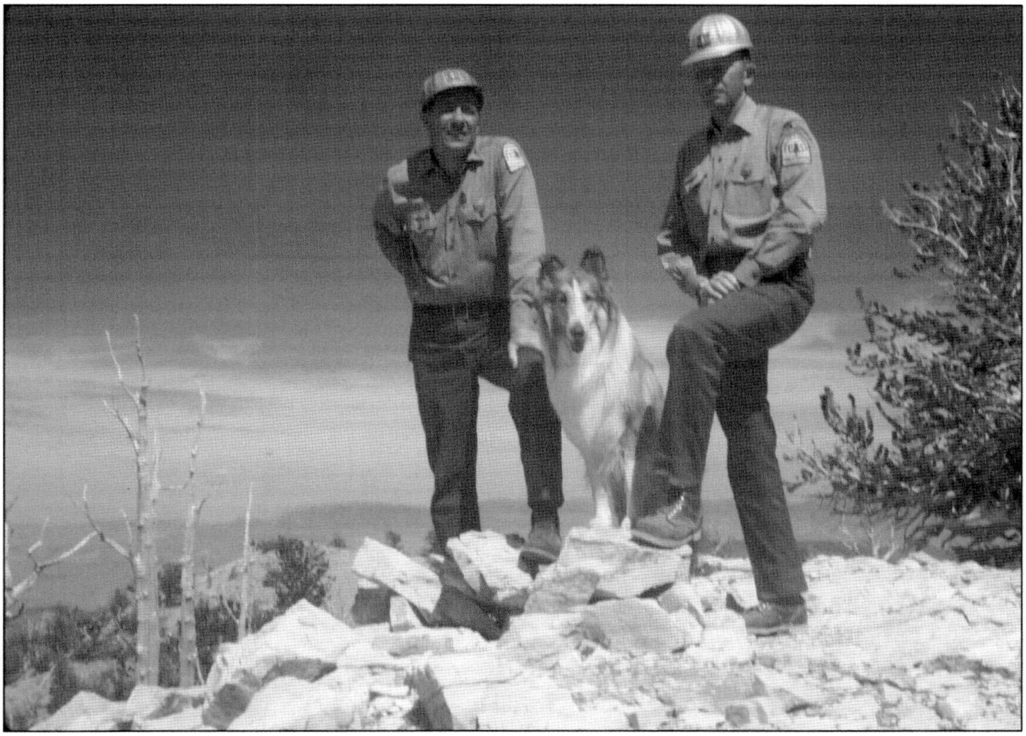

Actor Bob Bray (left), playing the role of Cory Stewart on the television show *Lassie*, poses with Lassie and an unidentified Forest Service liaison. In the early 1960s, at least three episodes of *Lassie* were filmed on the Inyo. One at Minaret Summit simulated an avalanche that buried someone, and Lassie rescued him. This 1964 episode was set in the Patriarch Grove of bristlecone pines. *Lassie* and other television shows helped give America a connection to public lands and a positive, if skin-deep, impression of the Forest Service. (Lee Prater photograph, courtesy of INF.)

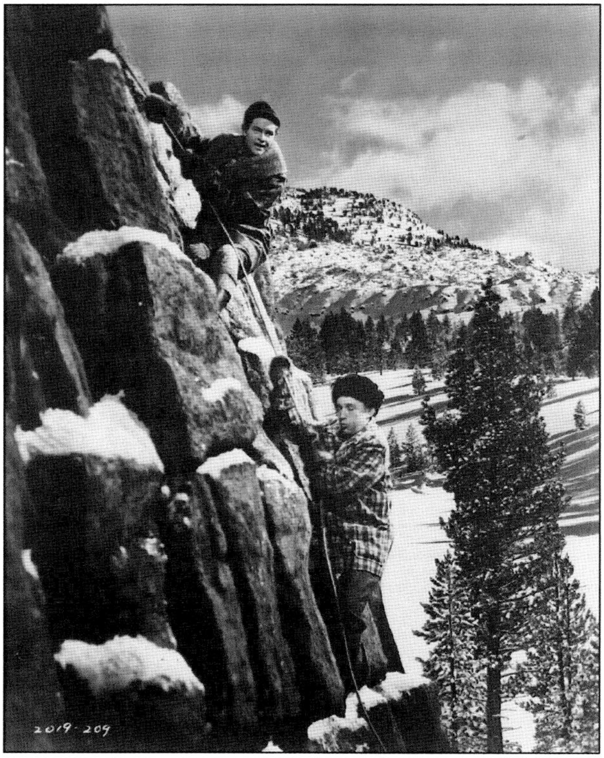

Bob Hope and Bing Crosby tackle a snowy cliff near Mammoth in the comedy *Road to Utopia* (1946). The pair played vaudeville actors who found a treasure map to an Alaskan gold mine, and the areas around Mammoth and June Lake provided the "Alaskan" environment. This was one of a series of "road to" movies that these two actors made. (Courtesy of Beverly and Jim Rogers Museum of Lone Pine Film History.)

Neesa (Janet Margolin) is an Indian maiden nursing the star of *Nevada Smith*, Steve McQueen, after gunslingers nearly drowned him in a river. Smith roams the West looking for his parents' murderers. His search begins nobly, but revenge turns Smith into a darker character. These scenes were filmed at Hot Creek in the upper Owens Valley. (Courtesy of Beverly and Jim Rogers Museum of Lone Pine Film History.)

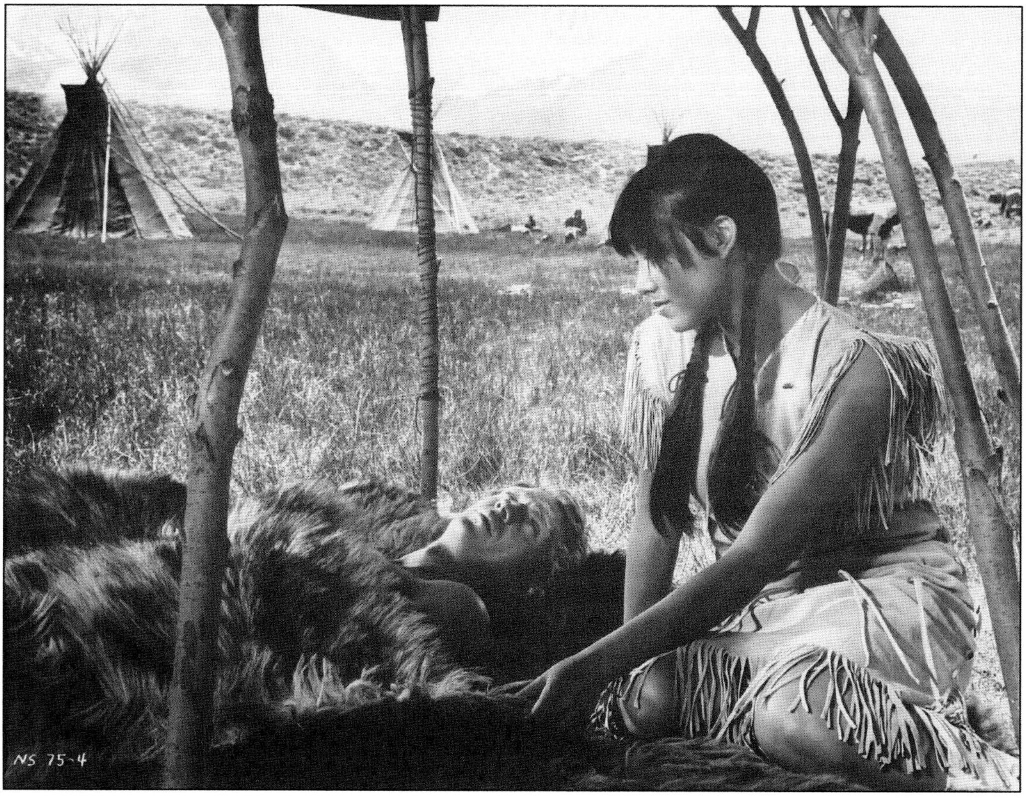

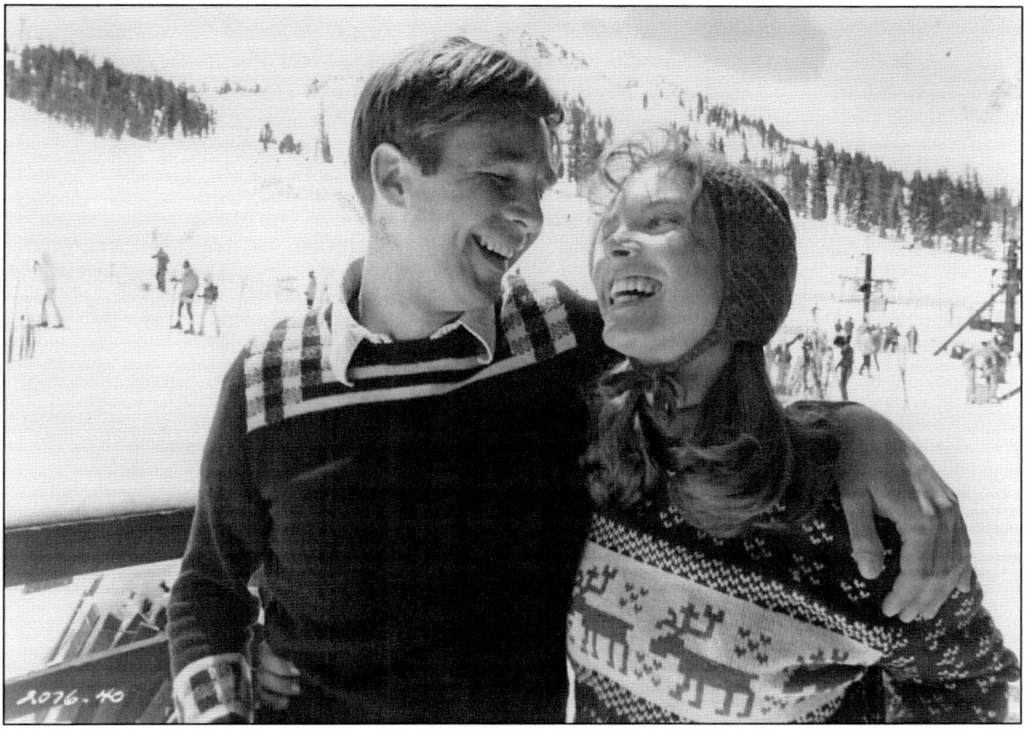

Beau Bridges and Marilyn Hassett played Dick Buek and Jill Kinmont in *The Other Side of the Mountain*, a 1975 docudrama about Kinmont's life. Born and raised in Bishop, Kinmont trained under Dave McCoy and became America's leading slalom racer. At the 1955 Olympic trials, she misgauged a bump and sailed off the run at high speed, breaking her neck and paralyzing her. *The Other Side of the Mountain* was filmed almost entirely in the Eastern Sierra. Kinmont says they did a reasonable job portraying her story, except it was more teary and sentimental than she. (Courtesy of Beverly and Jim Rogers Museum of Lone Pine Film History.)

Jill Kinmont developed mobility in her arms as best she could, got a teaching credential, and taught school for 32 years, including students with both physical and emotional disabilities. She founded and still manages the Jill Kinmont Indian Education Fund, which gives college scholarships to Native Americans. She is still a celebrity inspiration and gets "phone calls every week, from a person contemplating committing suicide to someone who is disabled, to a school kid doing a research paper." She also paints with watercolors, focusing on Eastern Sierra landscapes, the scenery she saw growing up on horses and skis and that she sees out of her window today. Here, she works on her back porch. (Courtesy of AS.)

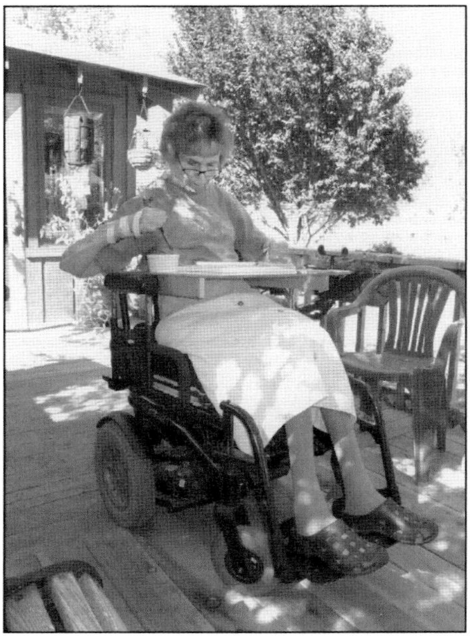

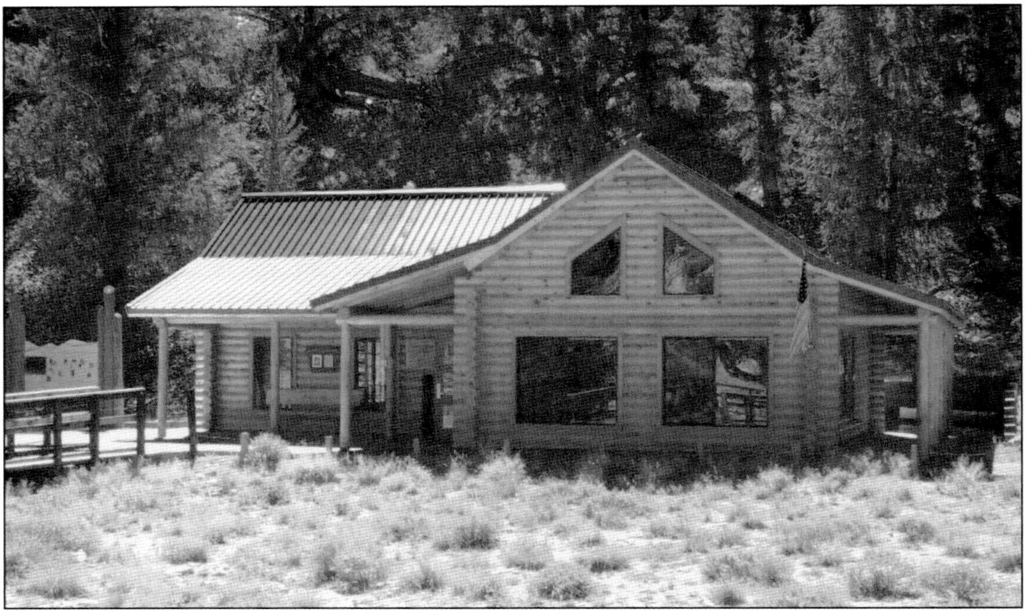

The Inyo built this visitor center at over 10,000 feet in the White Mountains for the Ancient Bristlecone Pine Forest in the 1990s. In September 2008, a deranged, antigovernment arsonist burned it down. The Eastern Sierra arts community gathered to raise money to rebuild a new center, and this sparked federal funding. A new structure is slated to open in 2012. Below is a photograph by John Dittli, one of the prominent local artists who contributed to the fund. (Above, courtesy of INF; below, © John Dittli.)

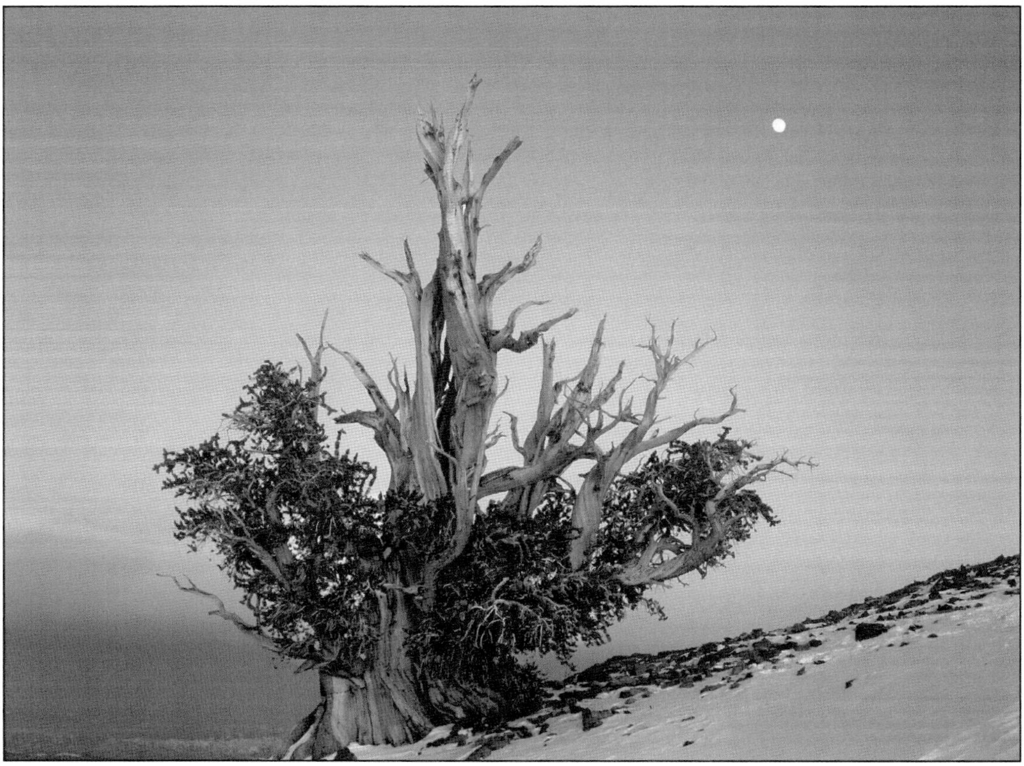

Tom Ross has made some 1,200 Sierra summit climbs, including all 248 peaks on the Sierra Club's list, and he always carried a camera. Tom also skied with a camera, took photographic overflights, and created some of the finest black-and-white images the Sierra has seen. Through the huge winter of 1969, his heart soared to see so much snow, and he took many photographs. Here is *The Minarets from San Joaquin Ridge* from that year. (© Tom Ross.)

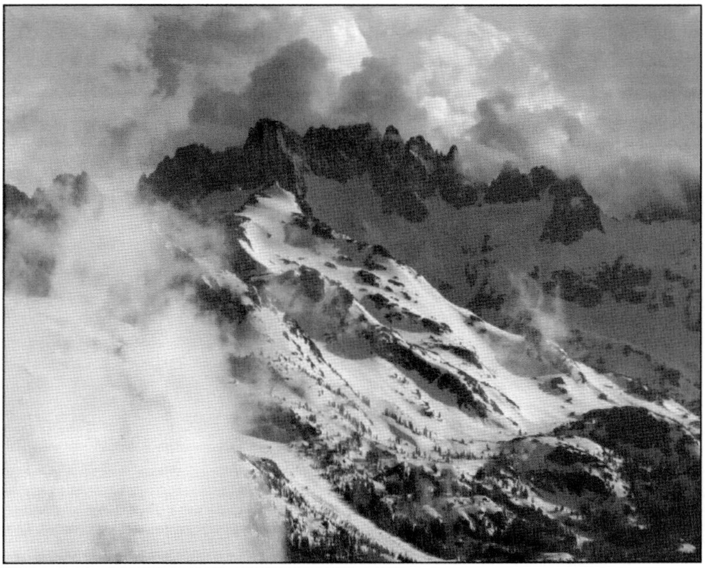

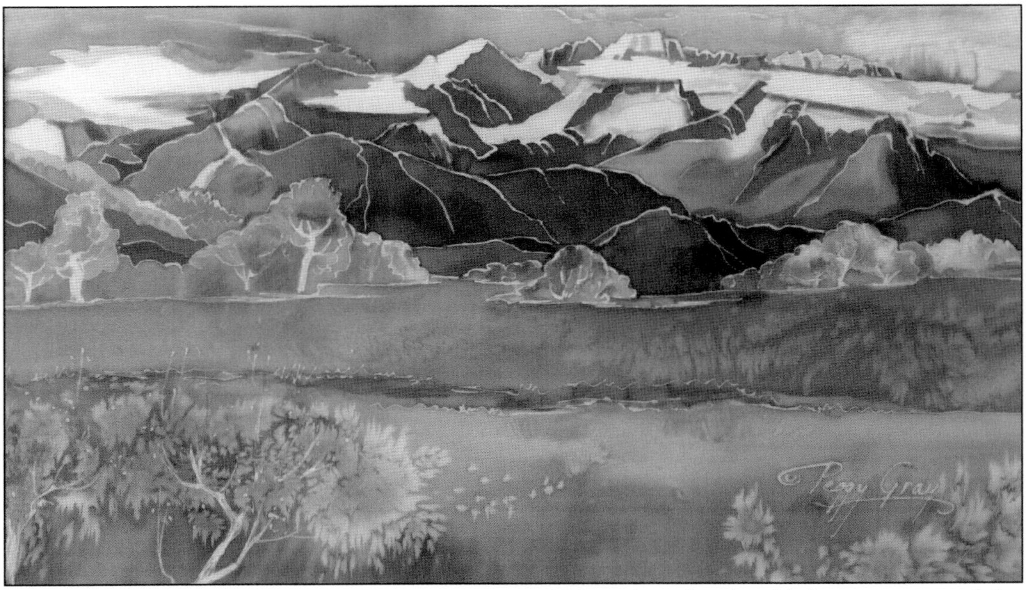

When Peggy Gray grew up in Korea, someone told her that she should draw pictures of the interesting places she had been. She did and never stopped. She got her master's in fine art from the University of California, Berkeley, and in 1954, moved to Bishop when her husband, Ray, was hired as an engineer at the Pine Creek Mine. She taught art at the high school and worked in watercolors. Gradually, she became the premier living painter in the Eastern Sierra. Here is one of her dynamic interpretations of the Sierra from near Bishop. (Courtesy of Chuck and Marin Spencer collection.)

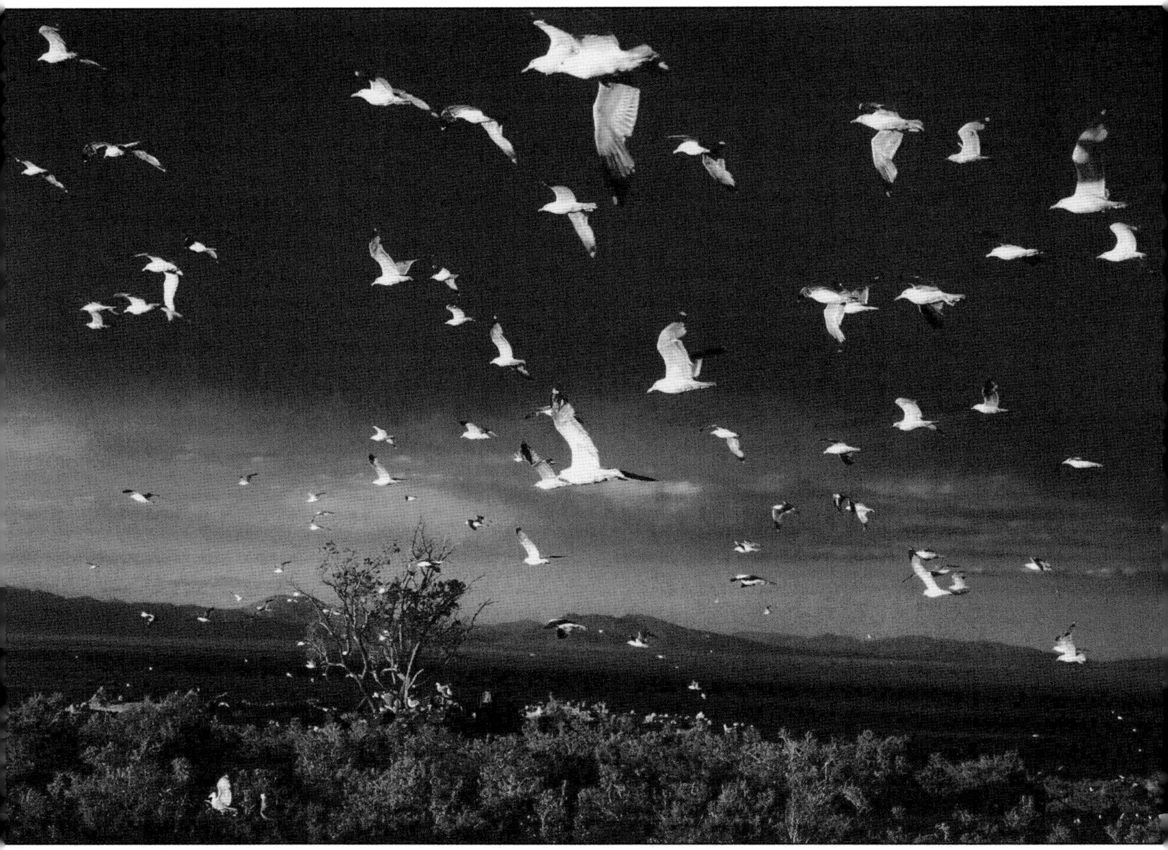

Galen Rowell learned in 1974 that Mono Lake was in trouble. With all inflow going to Los Angeles, the lake was dropping 20 inches per year and becoming poisonously saline. Scientists knew that the lake teemed with life (if not fish) that was doomed if nothing changed, but the public had barely heard of the place. In particular, thousands of California gulls nested at Negit Island, and their sanctuary was threatened. Rowell visited Negit and, as gulls filled the air, he laid down, letting them fly even closer. He took this shot of them sunlit against dark clouds and broke the Mono Lake story in *Audubon*. Rowell went on to invent modern adventure photography, defining his work as discovering visions of "dynamic landscape" arising out of immersion in spectacular places, light, and activities. After four decades of world travel, he and his wife, Barbara, settled in the Eastern Sierra. Their Bishop gallery remains a landmark even after their deaths in a 2002 plane crash. (Courtesy of Galen Rowell/Mountain Light.)

Gordon Wiltsie grew up in Bishop, where mentors including his high school photography teacher and Galen Rowell steered him to see the Sierra and Whites as grand ranges to explore. After college, travels to the Caucasus, Bangladesh, Nepal, and India gave Wiltsie a worldly sense for people as well as terrain. With hard work and his wife, Meredith, as foundations, he became one of America's elite adventure photographers, often working for *National Geographic*. Excursions to Antarctica particularly challenged, inspired, and boosted his career. After photographing worldwide, Wiltsie says that his roots in the Eastern Sierra directed him to seek places with "forces more powerful than anything human beings can create." Below, Wiltsie strings prayer flags on top of Temple Crag above Big Pine in honor of Smoke Blanchard. At right, his portrait of a bristlecone pine captures dynamic, ancient life illuminated by magical light. (Right, courtesy of Gordon Wiltsie; below, courtesy of AS.)

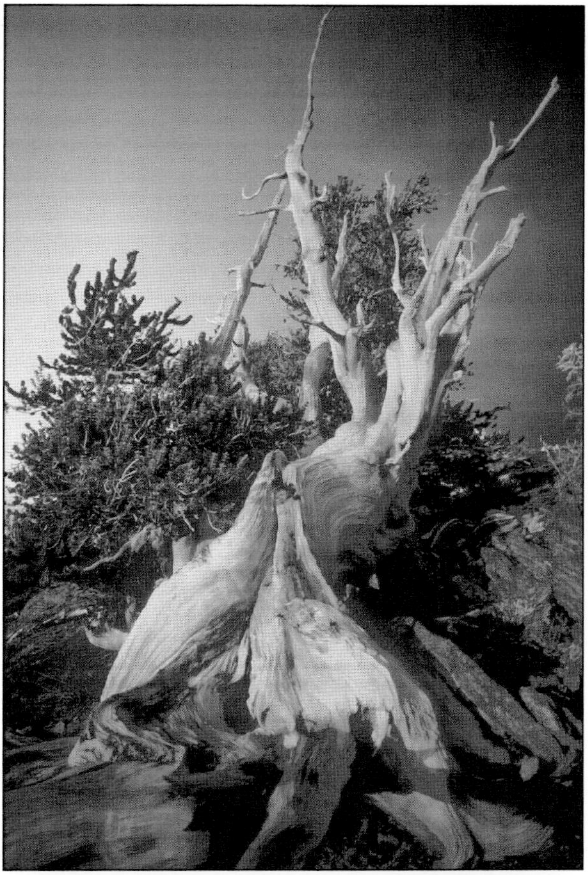

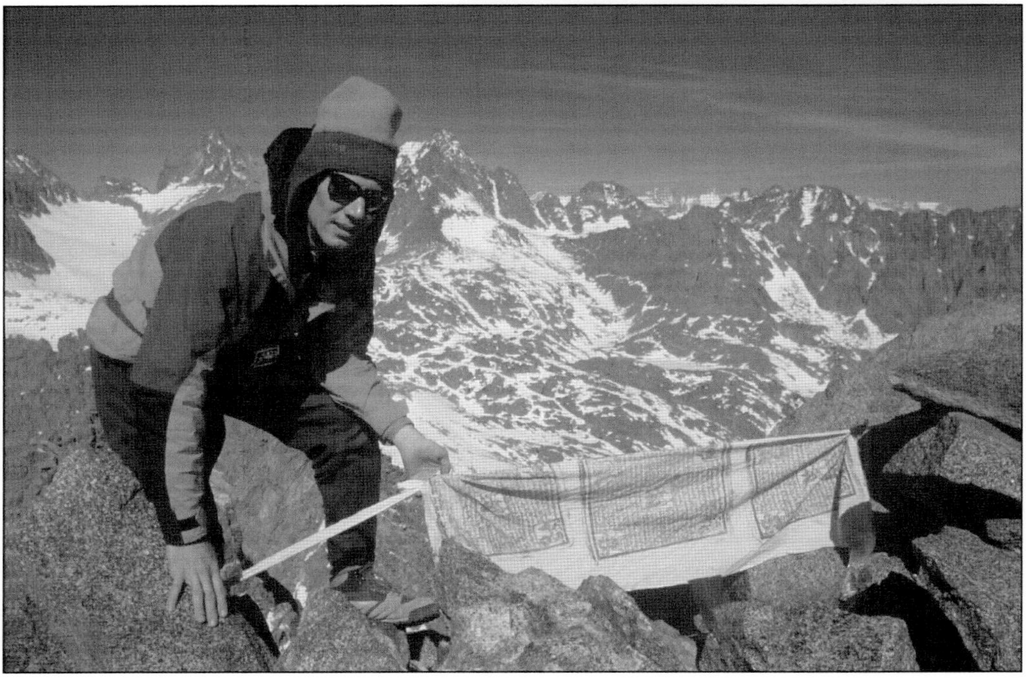

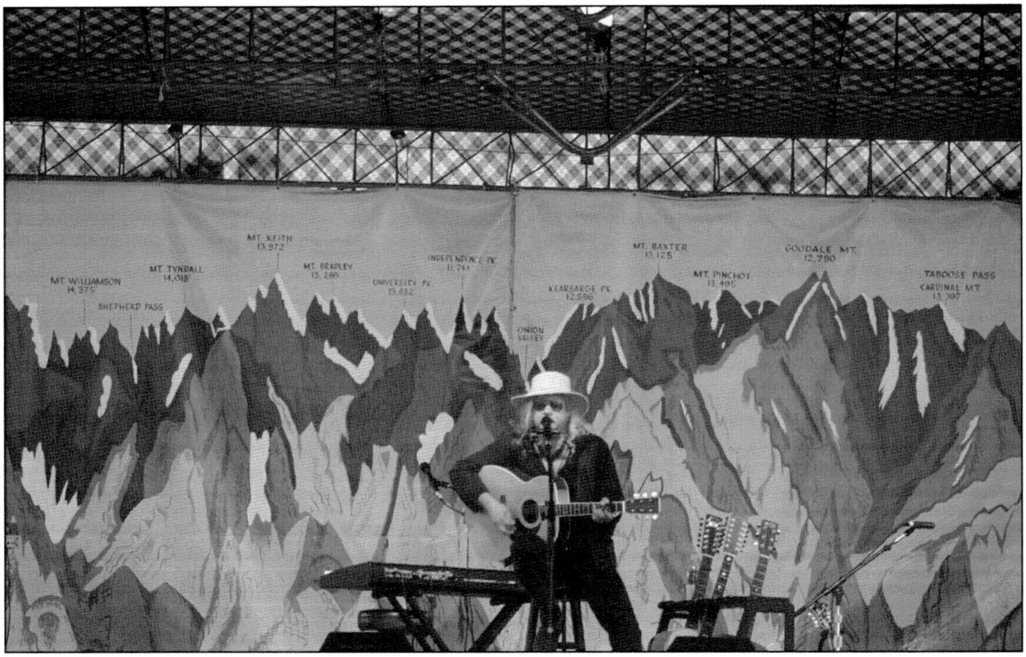

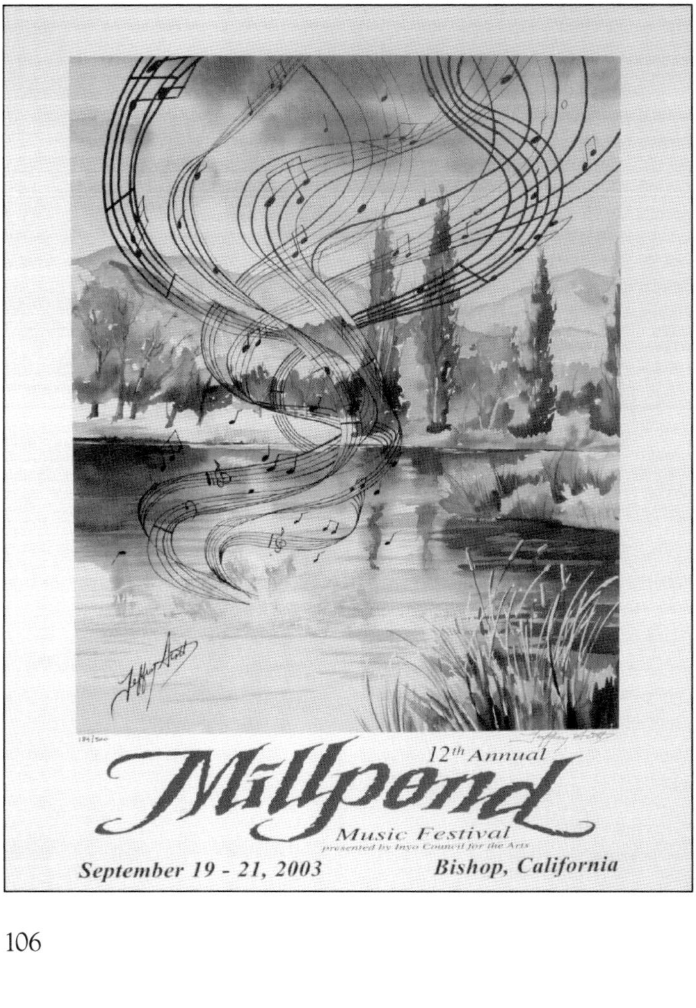

Arlo Guthrie captivates an audience at the Millpond Music Festival in 2004. The annual September festival is held at the county park that once was the site of the Inyo Lumber and Milling Company. This annual Inyo Council for the Arts festival draws notable musicians and an audience from far and wide, partly because people like the opportunity to visit the Eastern Sierra. Guthrie came because his father Woody's first wife, Mary Boyle, moved to Bishop and prompted him to come see the mountains. Mammoth also has several annual music festivals, the longest-running being the Sierra Summer Festival. Artist Jeffrey Scott created posters for many festivals, including this one for Millpond. (Courtesy of AS.)

Nine

How Does It Work in the Mountains?

Science and Conservation

How does the natural world operate, and how do people affect it? Scientists have been seeking answers on the Inyo at least since the 1890s, when Stanford president David Starr Jordan started examining the golden trout. As awareness of California's mountains has spread, scientists have eagerly sought to study here. Today, there is ever more being learned and asked here on topics from the shift of species with global warming to the temperature of space and the function of the mammalian heart.

Great pioneering studies began in 1911 and 1914, when Joseph Grinnell, director of Berkeley's Museum of Vertebrate Zoology, led teams to collect specimens and count the animals along two Sierra transects across the Whitney and Yosemite regions. The meticulous data they collected gives modern scientists background to compare how animal populations have changed.

The Inyo started formally recognizing the value of natural environments for scientific research in 1933, when one of the nation's first Research Natural Areas (RNA) was designated west of Saddlebag Lake. Named for Harvey Monroe Hall, a founder of the national RNA program, the 3,883-acre area is still a field station for the Carnegie Institute and others.

For some decades, the University of California has operated two institutions that facilitate a majority of the research on the Inyo: the White Mountain Research Station (WMRS) and the Sierra Nevada Aquatic Research Laboratory (SNARL). WMRS supports many kinds of research out of two facilities in the White Mountains, plus a base near the Owens River. SNARL has its base along Convict Creek.

Science is not just for academic enchantment; land managers need it to ground their decisions. The Forest Service's mission today includes planning that integrates ecosystem health and public demands. Managing fire and reducing fuel build-up, for instance, are now a complex priority. Most crucially, species need protection, and the Inyo ecosystems shelter several important ones. As the great biologist Aldo Leopold famously advised, "To keep every cog and wheel is the first precaution of intelligent tinkering." Schools from near and far, from elementary to the university level, know the Inyo as a premier field trip destination, where they can see, experience, and measure nature at work.

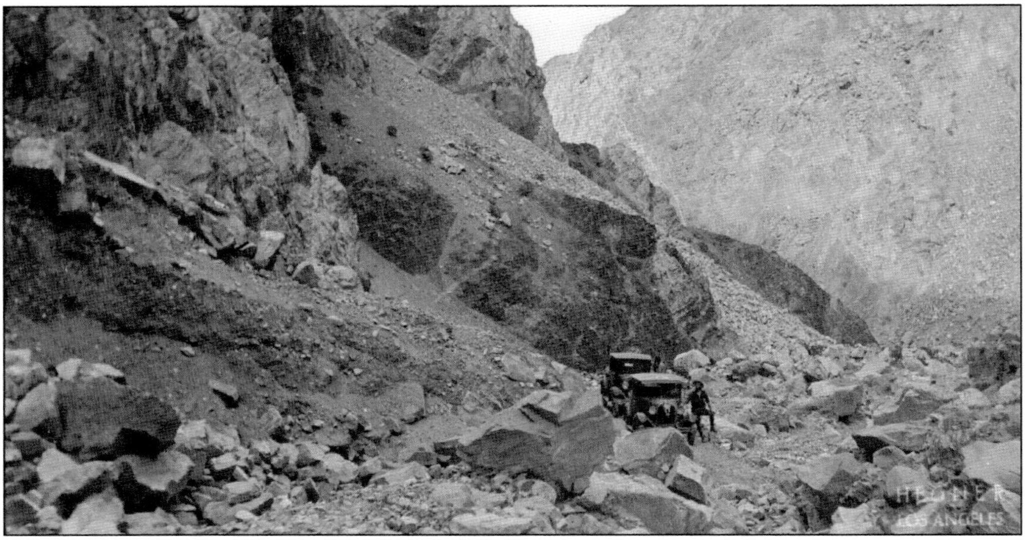

In 1931, Highland Park's Southwest Museum sent Clifford Park Baldwin on an archaeological expedition to Death Valley and the Inyo Mountains. Roads were rough, particularly this one through the Inyos to Saline Valley. As they explored into canyons on both sides of the range, Baldwin and his partners kept a substantial diary. Pictographs, petroglyphs, pottery shards, and mortars and pestles they found could have only inspired questions of how people managed in this hard, dry land, and how the climate might have been different in previous centuries. (Courtesy of ECM.)

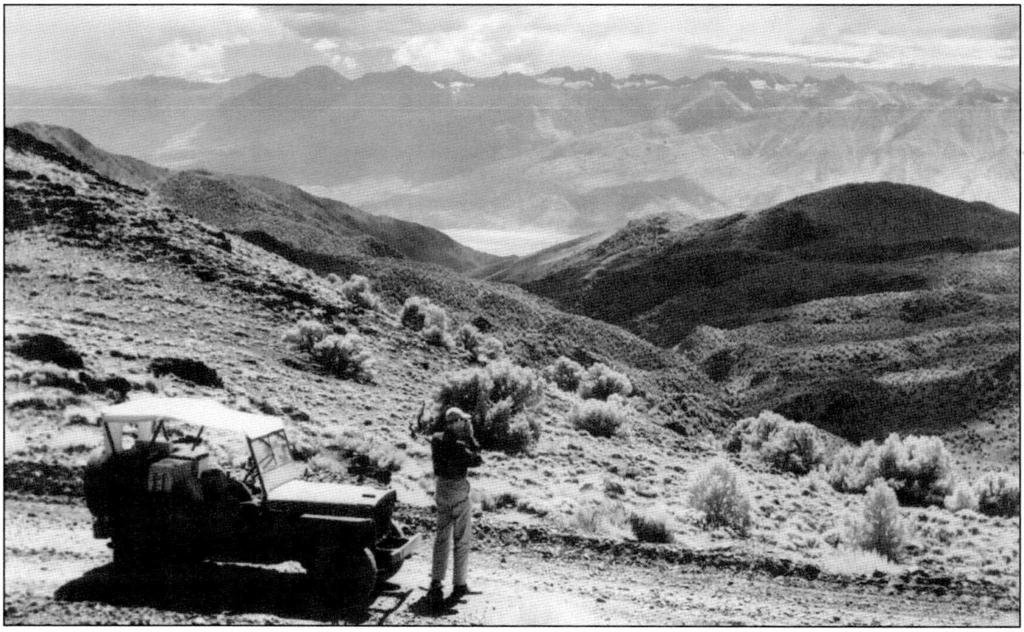

Dr. Nello Pace pauses to look at the Sierra from the White Mountains. Searching for a high-altitude location to study low-oxygen physiology, Pace founded the White Mountain Research Station in 1950. The US Navy, which was researching infrared missile sensors, offered to share the 10,200-foot-high facility at Crooked Creek (the old ranger station). Support from the National Science Foundation, the Rockefeller Foundation, and the University of California got the new research station going. (Courtesy of WMRS)

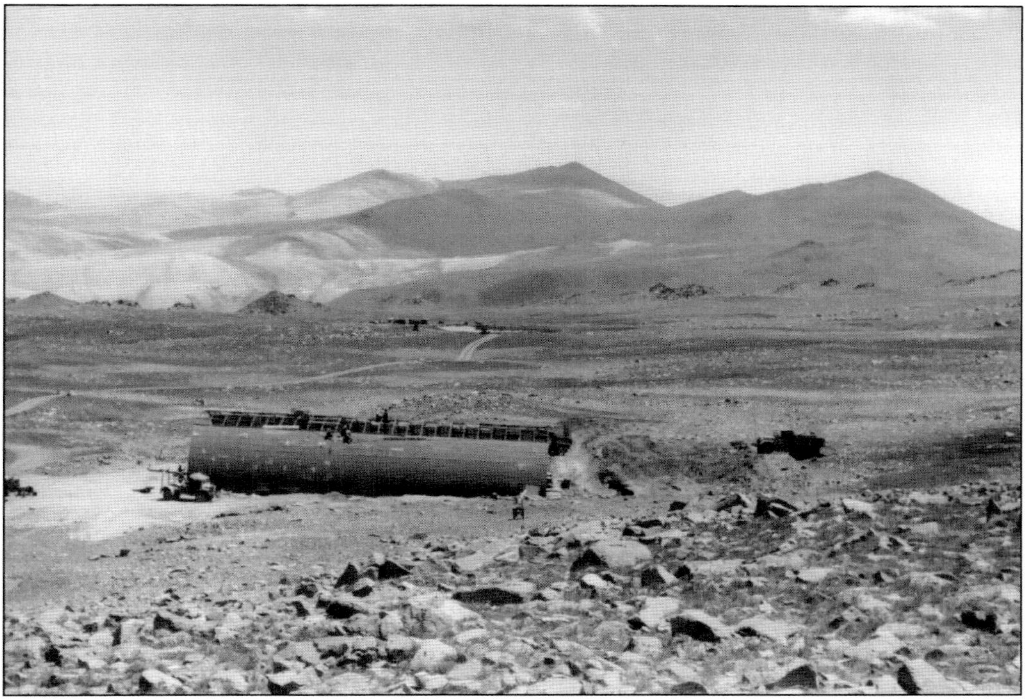

The White Mountain Research Station's initial goals included building one of the highest research facilities in North America. Here is the "Quonset Hut" at Mount Barcroft, elevation 12,500 feet on the crest of the White Mountains, under construction in 1951. The station has hosted over 2,000 studies and field trips. This site is famous for extreme winds—in fact, winds too strong for measuring devices. Below, in October 1950, soon after the opening of the initial White Mountain Research Station facilities at Crooked Creek, Jack Shriber sends a radio call to the University of California, Berkeley. (Both images courtesy of WMRS.)

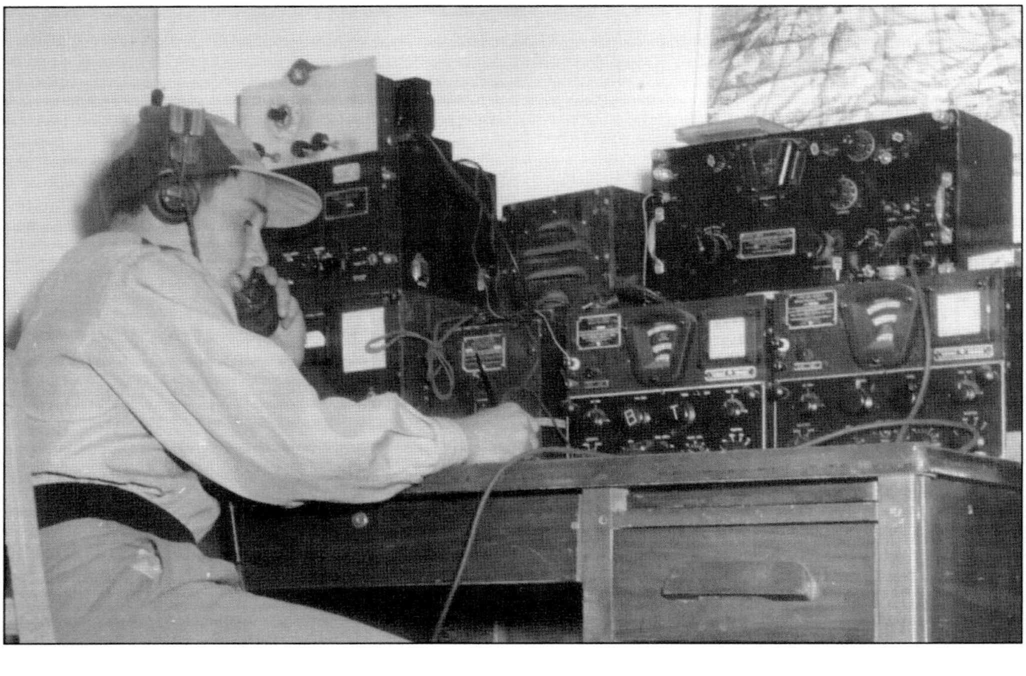

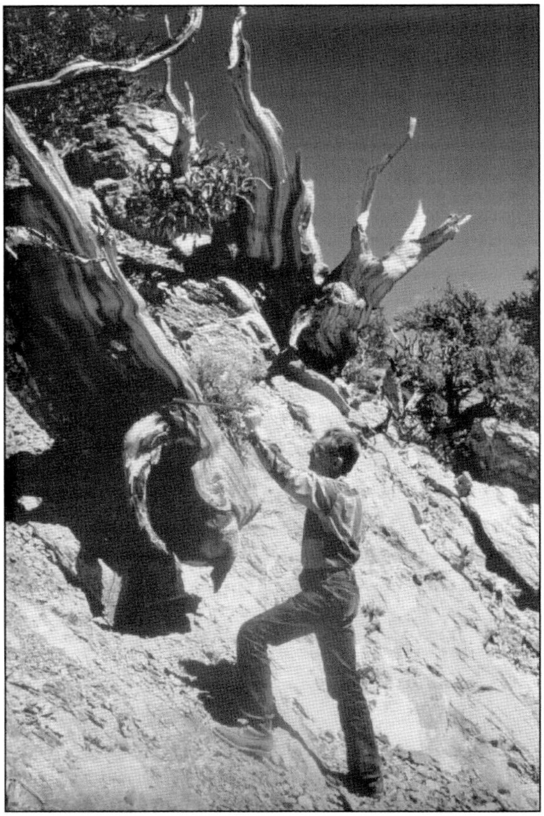

Around 1950, Dr. Ed Schulman of the University of Arizona began measuring growth rings in arid-zone trees, hoping to find evidence of past climates. He found ponderosa and piñon pines nearly 1,000 years old, showing interesting dry and wet periods. Inyo ranger A.E. Noren alerted Schulman to old trees in the White Mountains. In 1953, Schulman found trees over 3,500 years old. In 1957, he extracted cores from what is now known as the Methuselah Grove and found one tree with more than 4,600 annual rings. Since then, many old trees have been found, and the study of bristlecone rings is a powerful scientific tool. By measuring older, dead wood, the bristlecone record of climate now goes back some 12,000 years. At left, Schulman examines a bristlecone. Below, Schulman (left) and assistant M.E. "Spade" Cooley carefully measure the miniscule rings on bristlecone corings. Each core is mounted in a wooden trough for archival storage and study. (Courtesy of INF.)

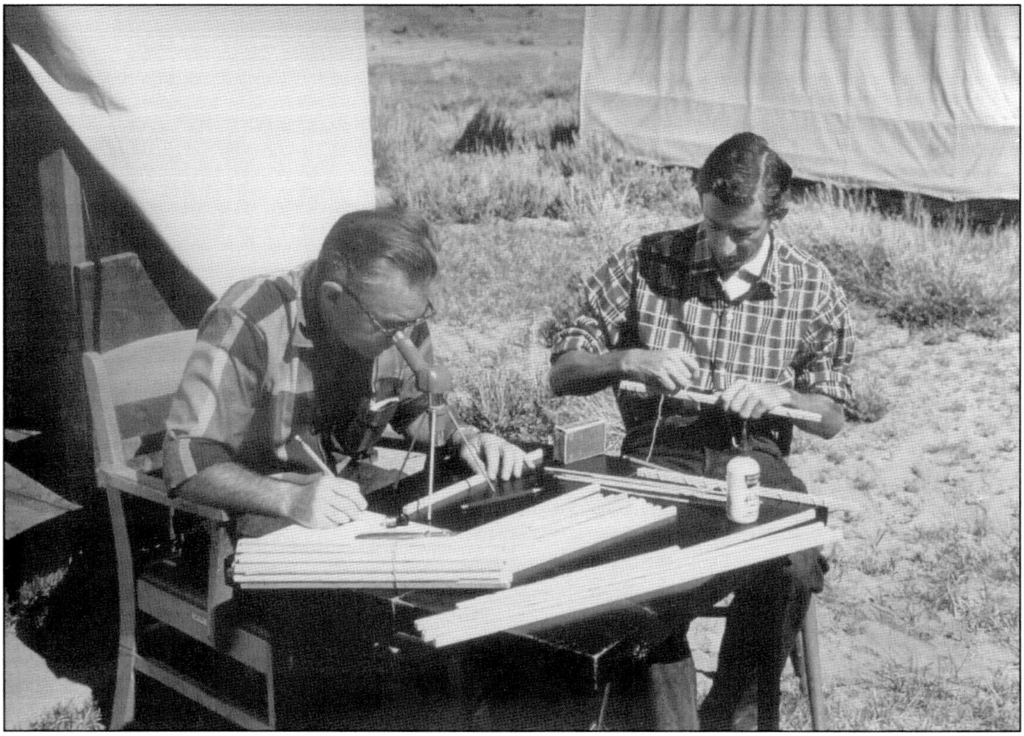

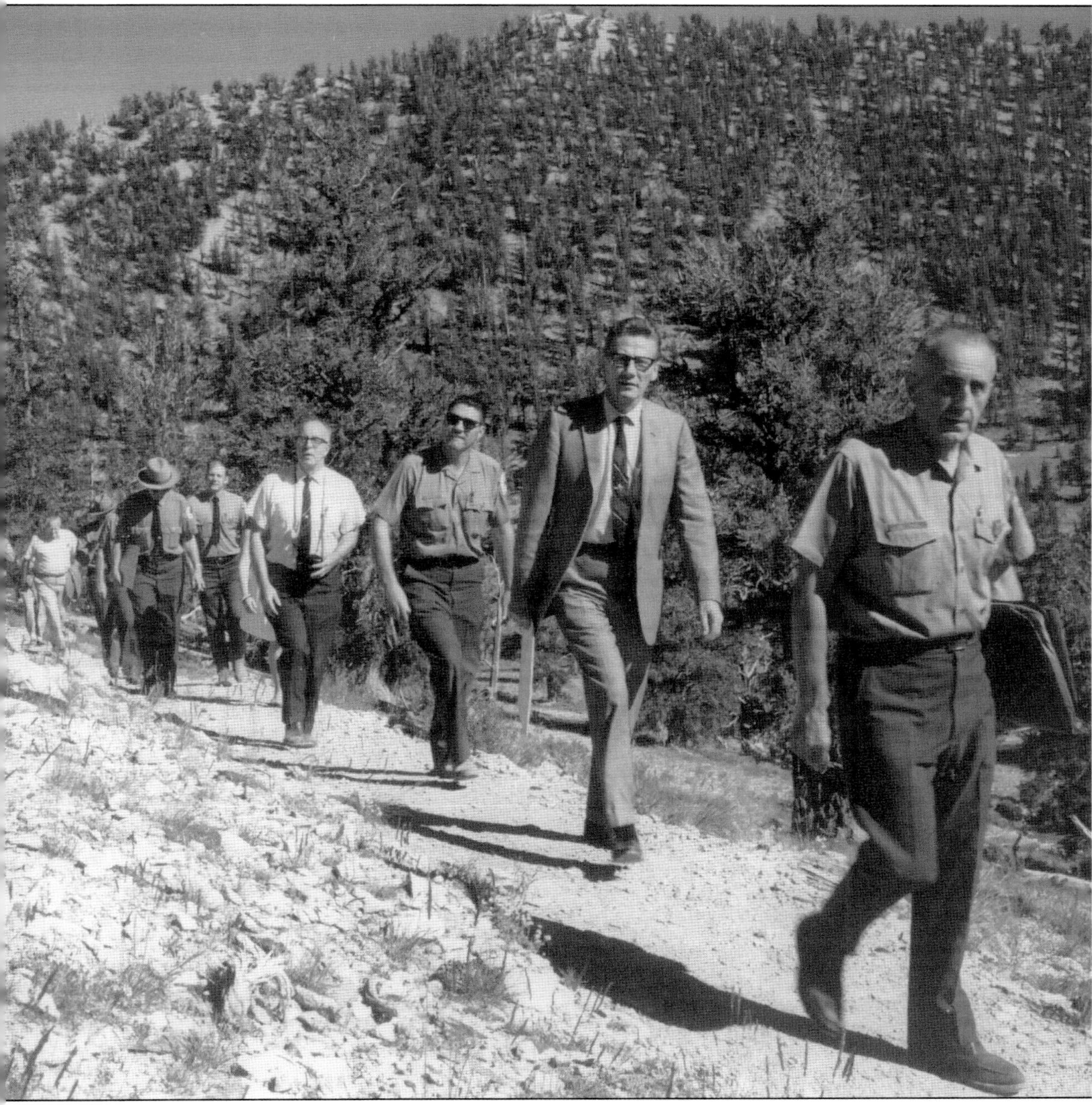

A year after Schulman's discoveries, the Forest Service designated the Ancient Bristlecone Pine Forest. As news spread that the White's bristlecones are the oldest known living things, people called for the Inyo to protect them. Famed scientist Linus Pauling declared that the groves should be fenced off and sequestered for scientists to explore. Many thought groves should be turned over to the National Park Service. Supervisor Radel declared that the Inyo would protect the ancient trees and keep them available for the public. In 1969, he convened a Bristlecone Conference, and on July 25, conferees toured the Grove. From right to left are Supervisor Radel; Assistant Secretary of Agriculture E.J. "Fritz" Behrens; White Mountain district ranger Hal McElroy; Chief Ed Cliff; Regional Officer Grant Morse; Ron McCormick, Inyo recreation officer; Regional Forester Jack Deinema; Congressman Harold "Bizz" Johnson; and Dick Harris, Inyo naturalist. The Inyo's management has been successful; today, the Forest maintains a visitor center and trails, keeping the identity of the oldest trees secret to prevent desecration. (Rocky Rockwell photograph, courtesy of INF.)

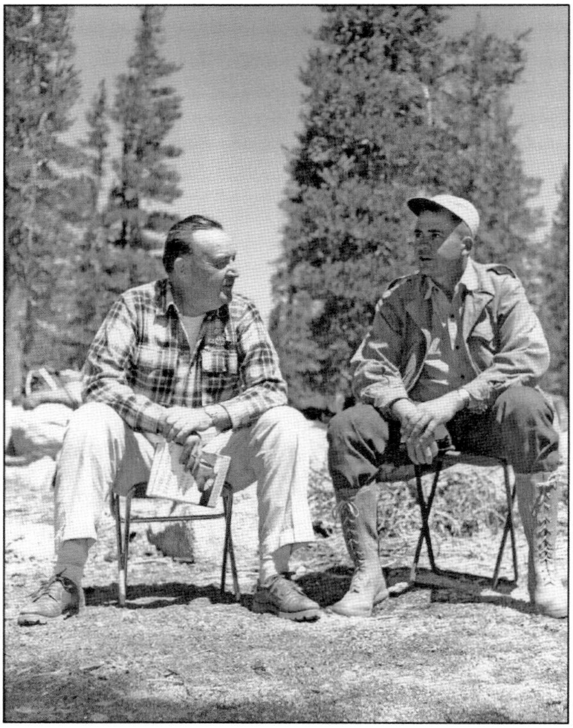

Gov. Pat Brown (left) chats with fisheries biologist Phil Pister at Big Whitney Meadow in August 1959. The governor's horseback trip to this remote setting was a logistical challenge, but it gave him a chance to see the golden trout—California's state fish since 1947—to meet with Fish and Game staff, and to publicize the Kern Plateau and the trout. A decade later, Pister coordinated an emergency program to save the golden trout from extinction. (Courtesy of Phil Pister collection.)

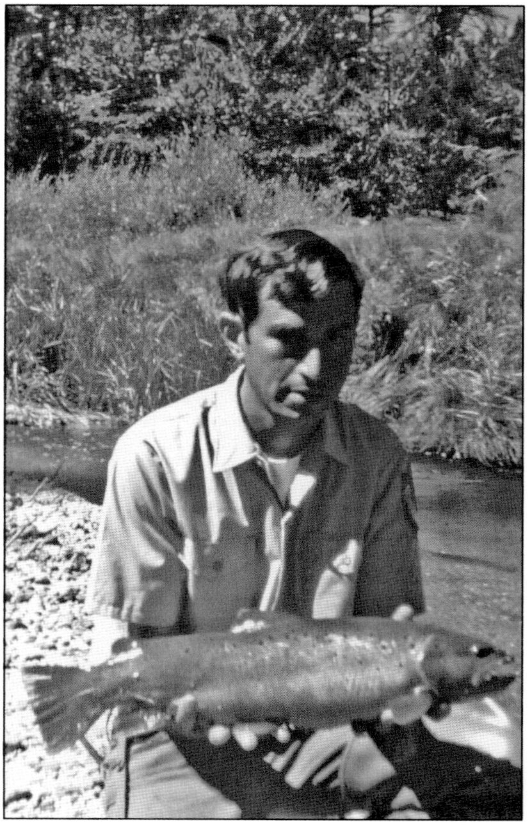

Fish and Game biologist Howard Shainberg holds a large brown trout taken from the South Fork Kern River in September 1969. Someone had planted European brown trout in the upper Kern drainage, and the fish had somehow migrated above Ramshaw Falls and were eating golden trout in their native sanctuary. Pister and Shainberg found 100 browns for every golden. Downstream, they found that planted rainbows were interbreeding with goldens, and they knew that hybridization was an even more insidious threat to the golden trout species. The California Department of Fish and Game then coordinated with the Inyo on an emergency plan to remove the brown trout and build enhanced barriers to block the exotics below. (Courtesy of California Department of Fish and Game.)

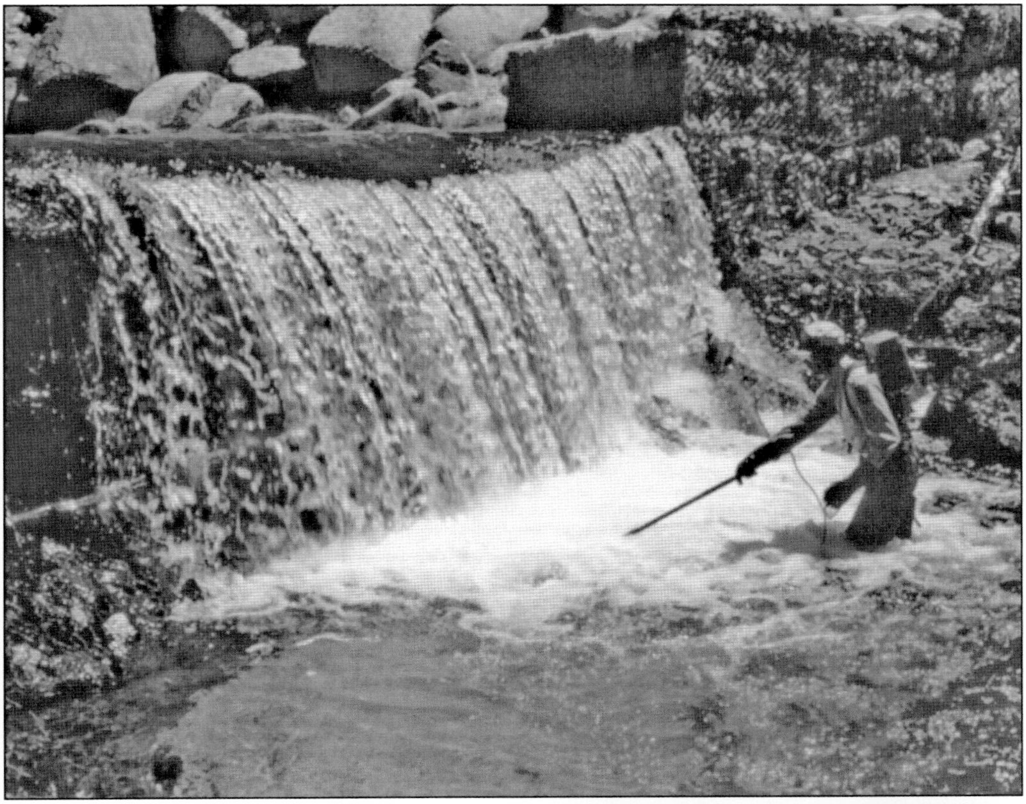

Fish and Game biologist Darrell Wong is shown above in 1985 electro-shocking trout below a temporary barrier. At right, Department of Fish and Game biologist Don Sada arranges a drip system of antimycin to kill fish in the upper reaches of the South Fork Kern River in October 1977. Luckily, Golden Trout Creek comes within 200 yards of this stream, so before the river was poisoned, biologists were able to transfer 6,000 golden trout in cages there and replace them after the poison was gone. The barriers, removals, and replacements basically worked, and the golden trout survives. The Forest Service lists it as a sensitive species, secure in an estimated four percent of its original range. (Both images courtesy of California Department of Fish and Game.)

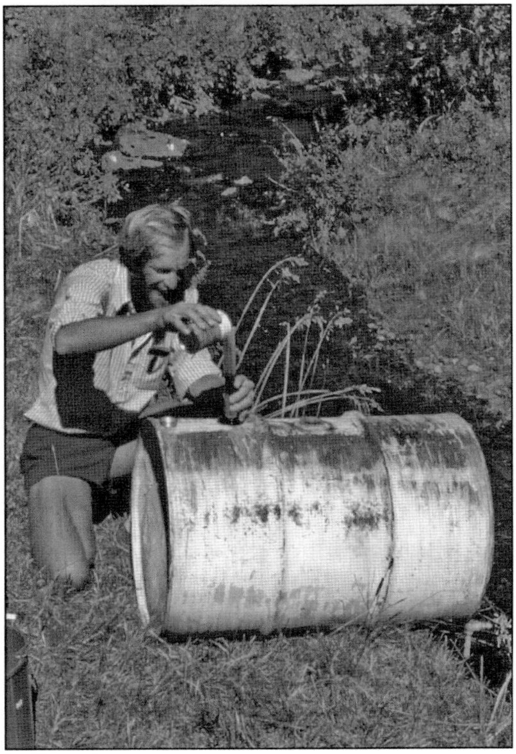

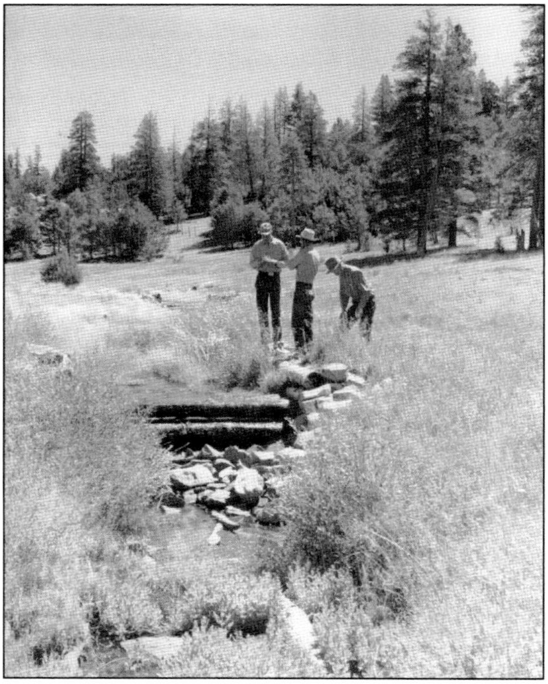

Cattle grazing on the Kern Plateau has been another problem for golden trout. Extensive, unmanaged grazing a century ago was especially damaging. Today, the Inyo works to restore the ecosystem and balance grazing with the needs of trout. Here, biologists inspect a silt barrier constructed at Casa Vieja Meadows, an area where golden trout were introduced. Grazing here had broken down stream banks and reduced vegetation, creating a stream that cut deep and ran fast, with a load of silt. Barriers such as this slow a stream down, settle silt, and stabilize banks, allowing vegetation to regrow and trout to survive. (Courtesy of INF.)

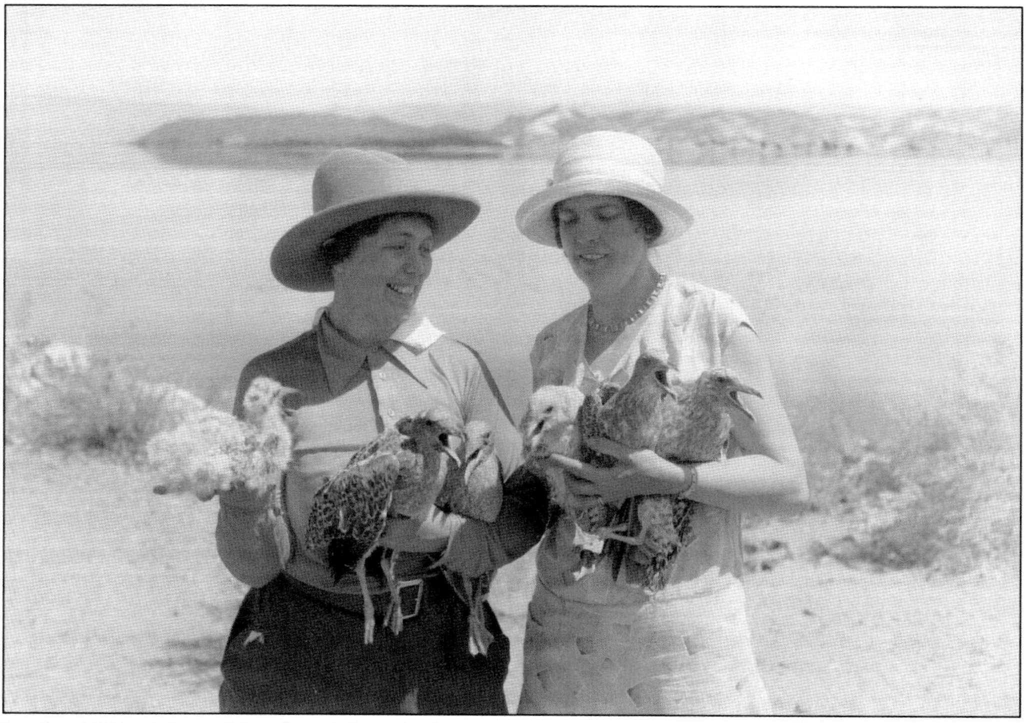

In the 1920s, Negit Island in Mono Lake was a tourist attraction as well as one of the most important nesting sites for California gulls. By 1980, the scenario that scientists worried about for Mono Lake came true. Los Angeles's inflow diversions (started in 1941) dropped the lake enough to expose a land bridge from the island to the mainland. Coyotes came to the island to eat gulls, the gull parents fled, and their chicks starved. (Courtesy of ECM.)

David Gaines, cofounder with his wife, Sally, of the Mono Lake Committee, stands next to a pile of dead gull chicks on Negit Island in 1981. The Negit land bridge spurred the committee to redouble its public outreach and legal efforts. The Audubon Society joined its lawsuit, and in 1983, the California Supreme Court ruled that the public trust values of Mono Lake take precedence over previous water agreements and that Los Angeles would have to maintain the lake at a viable level. (Courtesy of Mono Lake Committee.)

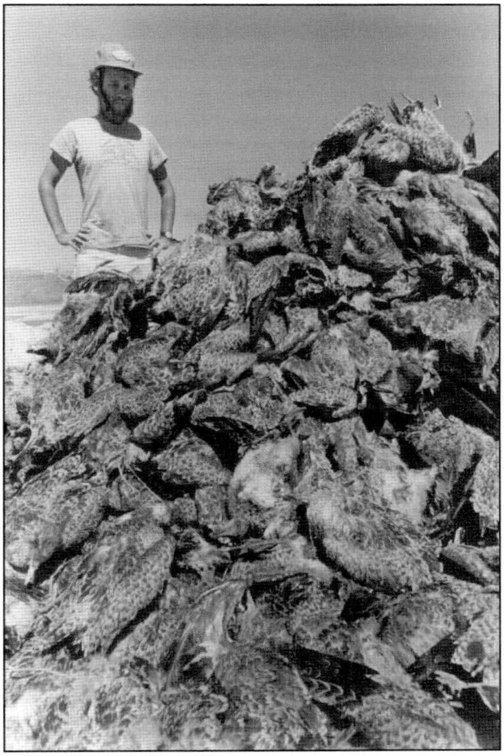

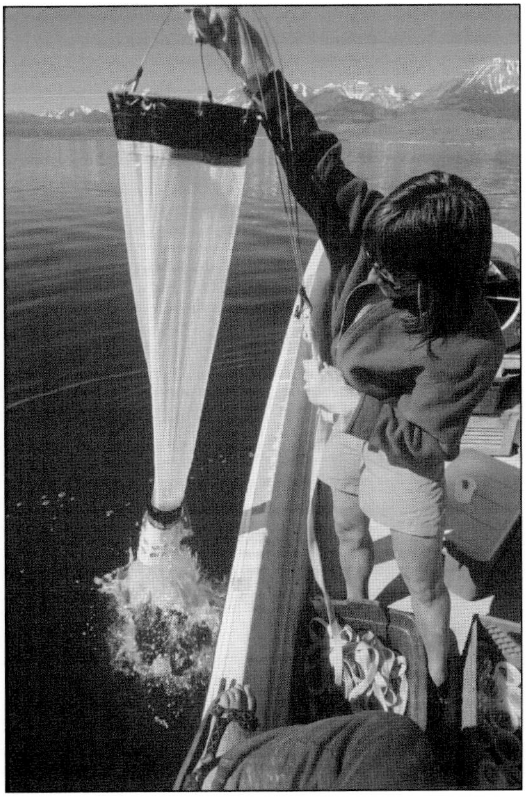

Research assistant Darla Heil hauls a sample of water, algae, and brine shrimp from Mono Lake in 1991. She worked under Dr. Robert Jellison through the Sierra Nevada Aquatic Research Laboratory. His and other research focused on levels of salinity and the health of the brine shrimp and algae, the basis of the ecosystem. Based on such research, the California Water Resources Control Board decided in 1994 that Los Angeles would have to maintain Mono Lake at 6,392 feet elevation, 25 feet below the level before diversions and 17 feet above the 1982 low point. (Courtesy of AS.)

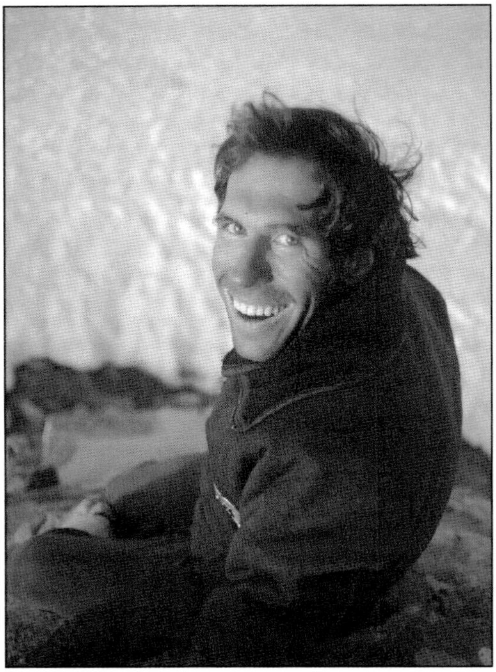

Walter Rosenthal was an extraordinary climber and skier; in the late 1970s and early 1980s, he pioneered first descents of some of the steepest ski runs in the Eastern Sierra. In 1991, he turned to snow research and worked through the Institute for Computational Earth System Science at the University of California, Santa Barbara. From assessing snow at the molecular level to measuring snowpacks from satellite, Rosenthal started unraveling deeper truths about snow. By 2005, he was starting to redefine how snow crystals "sinter" together and stabilize a snowpack. He also founded the Eastern Sierra Avalanche Center, a nonprofit to coordinate a backcountry-forecasting program with the Inyo. Sadly, he was killed in 2006 trying to save fellow ski patrolmen from a cave-in at a volcanic vent on Mammoth Mountain. (Courtesy of Jim Stimson.)

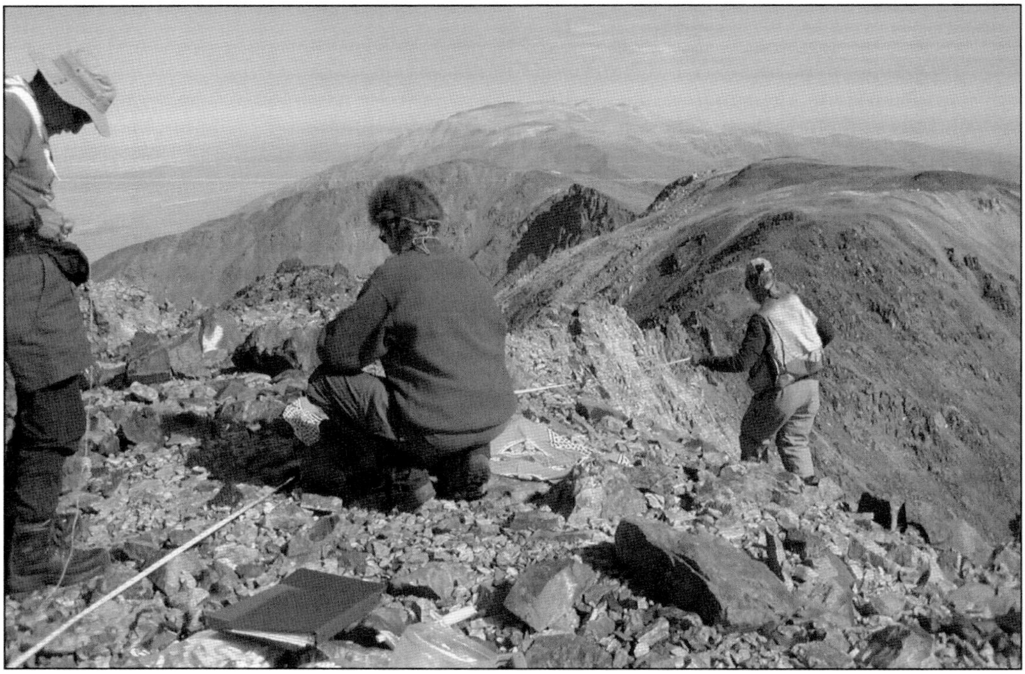

Dr. Connie Millar (center) works with two assistants atop the White Mountains in 2009 for an international study called GLORIA, Global Observation Research Initiative in Alpine Environments. WMRS supports this long-term study locally. It began in 2004 by precisely documenting plant life at selected plots. The plots are to be revisited again and again, building a worldwide measure of how high-altitude life and conditions might be changing, particularly with global warming. Dr. Millar is an expert in assessing ancient climates and is employed by the Forest Service's Pacific Southwest Research Station. (Jim Bishop photograph, courtesy of WMRS.)

Dr. John Wehausen began following Sierra bighorn sheep in 1974. Traversing canyons and mountainsides, he learned all about the lifestyles of the bighorn. In the 1980s, he started noticing fewer bighorn and how, in winter, they were starving up high instead of coming down to browse at the foot of the Sierra. Wehausen figured that, while diseases from domestic sheep had probably decimated bighorn from prehistoric populations, the sheep were now being attacked by mountain lions, another protected species. By 1995, he counted only about 100 bighorn in the whole Sierra—less than half the 1975 population. In 1999, the US Fish and Wildlife Service listed Sierra bighorn as endangered, and thus, the California Department of Fish and Game has been able to kill the occasional mountain lion preying on bighorn. Since then, the sheep have recovered to a population of over 350. At right, Wehausen looks for bighorn in Pine Creek. Below is a group of rams transplanted into Lee Vining Canyon. (Both images courtesy of AS.)

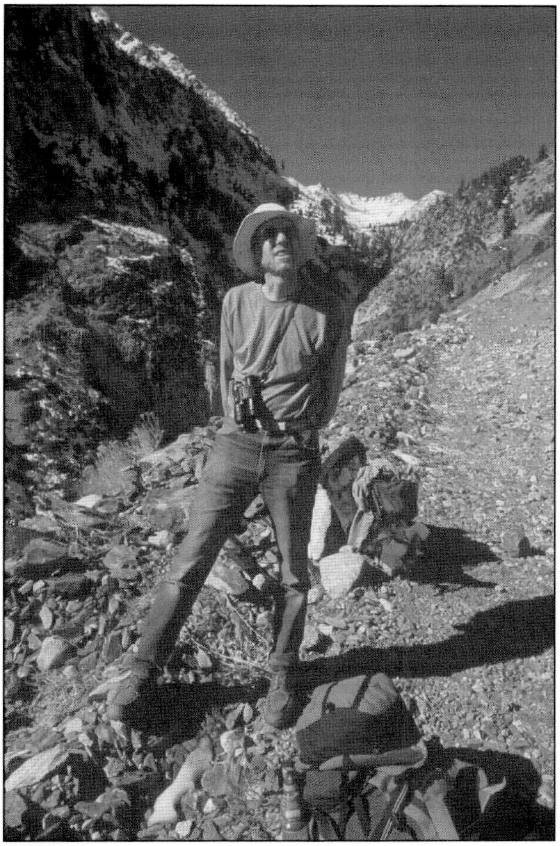

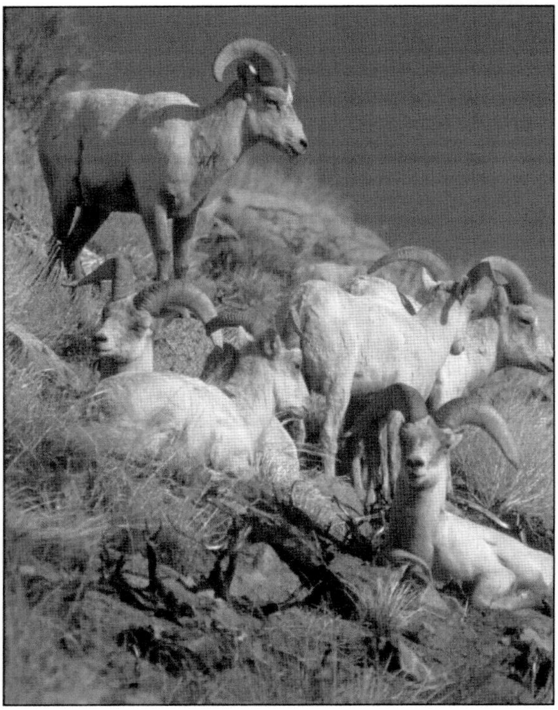

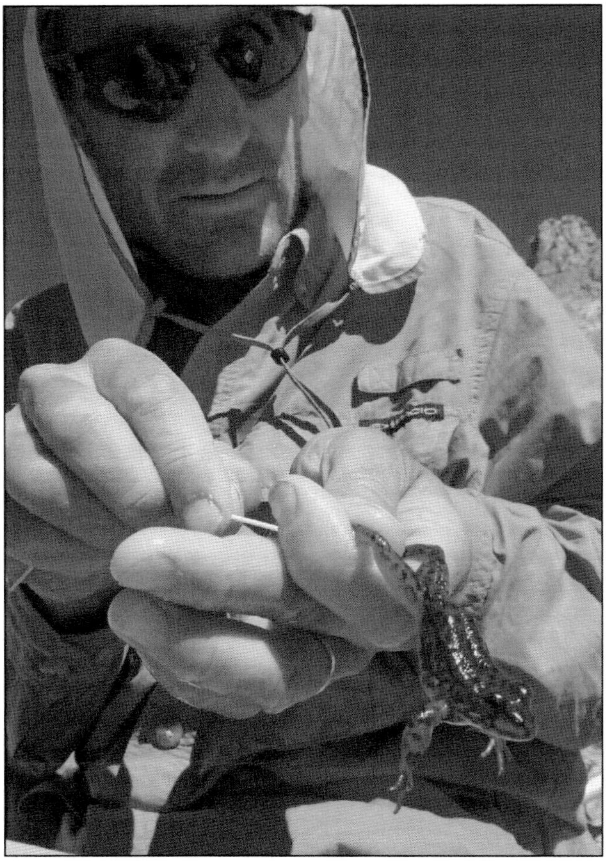

The pre–World War I surveys of Joseph Grinnell told of Sierra lakes teeming with an unusual amphibian of frigid waters—the mountain yellow-legged frog. When Grinnell walked along a tarn at 11,000 feet, frogs and tadpoles splashed at the water's edge; they were the most common vertebrates above timberline. In the 1990s, Dr. Roland Knapp and others surveyed over 15,000 lakes and creeks and found only a few remaining populations. Careful study confirmed Grinnell's observation that planted trout eat the frogs at these lakes. Knapp and the CDFG worked out a plan to remove trout from some less-viable fishing lakes and reestablish the frogs. The plan worked; the upper two Big Pine Lakes and an upper Treasure Lake saw the frogs return in glorious frenzy. Unfortunately, a worldwide chytrid fungus has diminished some of the new populations. At left, Knapp swabs a mountain yellow-legged frog to test for the disease. (Both images courtesy of Roland Knapp.)

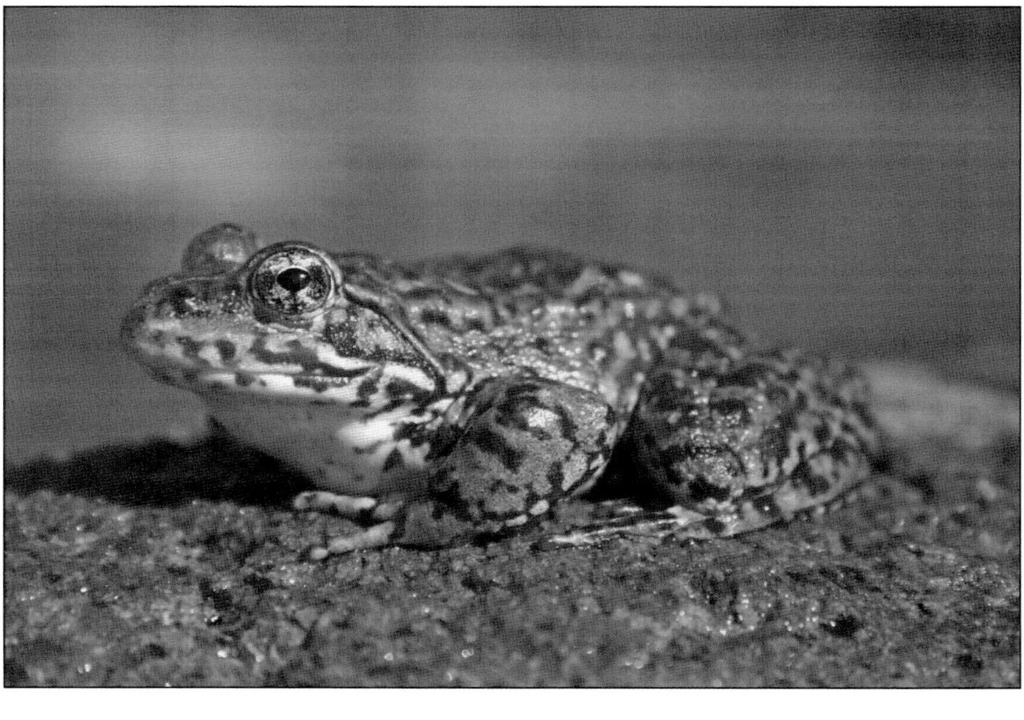

Ten

For the Enjoyment of All

Recreation

Recreation is important. People spend millions of hours and dollars on entertainment and play, yet do not always know to appreciate the depth it can bring—particularly recreation in the outdoors. As the population has evolved into a complex society, where bodies and minds are easily alienated from exercise and nature, outdoor recreation becomes a primary channel where minds can rest in open space and bodies can exercise in confluence with terrain. People are drawn to the Inyo because it is magnificent land; with mindful play, visitors find that the land draws them beyond their own creations and into a level of re-creation as people who know a bit more of what the broader community of life is about.

Today, outdoor recreation has evolved into dozens of activity spectra, from snowboarding to trail running and from snowmobiling to kite boarding. Skiing alone now has at least six subcategories. All this is spectacularly good, but it is even better when people remember to hold focus to the foundation of it all, the landscape. Recreation is best when it opens people's eyes, ears, noses, skin, and minds to the world of wild creation.

The very existence of the Inyo National Forest—magnificent terrain managed conservatively and mostly for public water supply and recreation—reminds people that when they go into a landscape, it is they who will gain from adapting to the land. Knowing their position on the map and role in the terrain are seeds to having at-home experiences in the landscapes they visit. Two great luminaries have pointed the way. Aldo Leopold wrote, "Recreational development is a job not of building roads into lovely country, but of building receptivity into the still unlovely human mind." And of course, John Muir penned, "Going to the mountains is going home."

From the very establishment of the Inyo National Forest to the designations of its nine wilderness areas, two special designation areas, seven Research Natural Areas, and the ever-evolving iterations of the Inyo's Forest Plan, America has built institutional respect for these mountains. Few places in the world are so blessed with such an amazing diversity of nature, and Americans can be proud to call it theirs to enjoy.

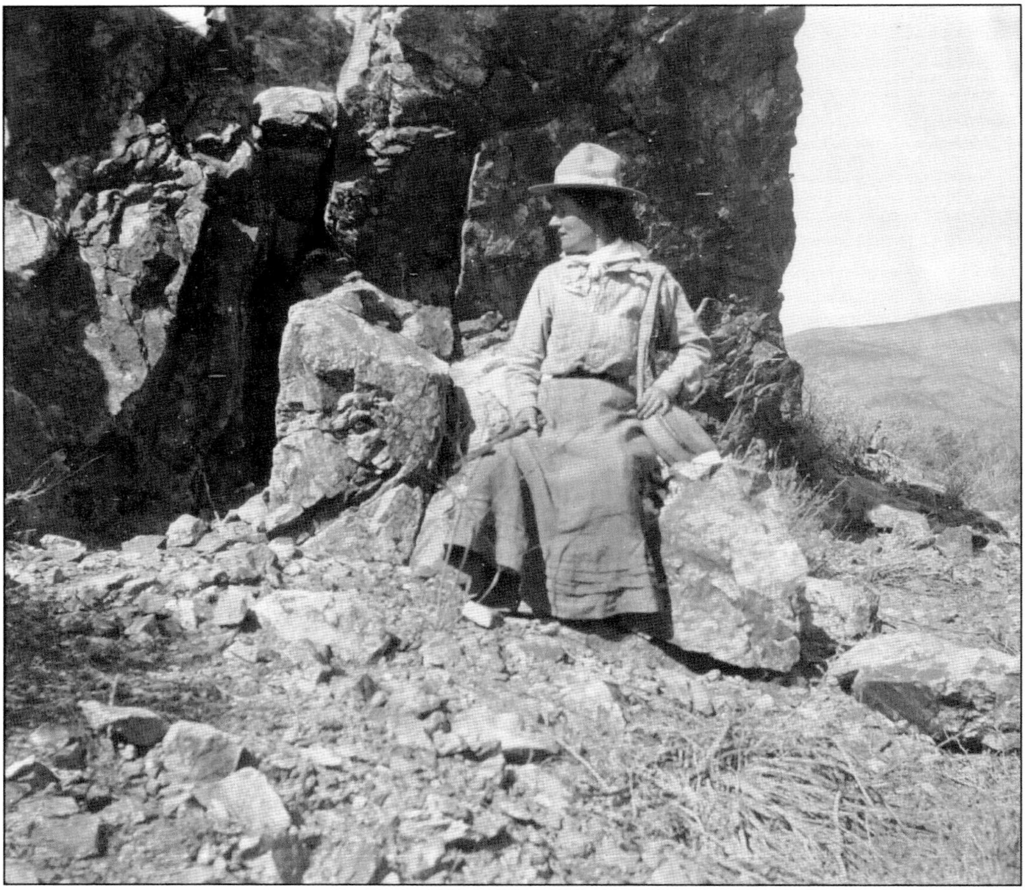

Around the end of the 19th century, America became more sedentary, and the country started to see outdoor exercise not as toil but a healthful necessity. Here, a woman of the early 20th century takes a rest on a hike on an unknown location on the Inyo. (Courtesy of ECM.)

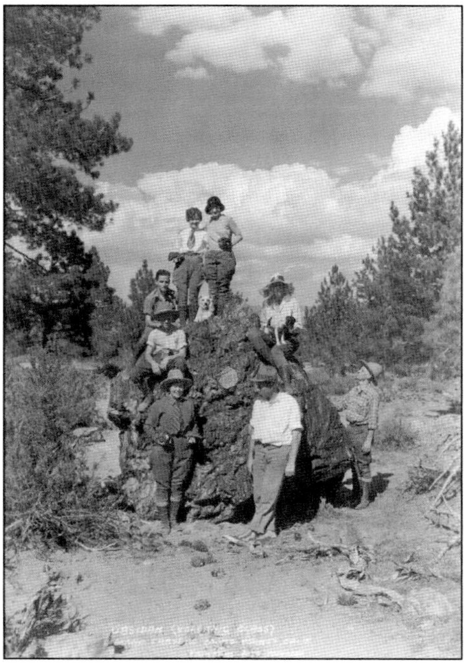

The Inyo holds the premier obsidian fields in North America. South of June Lake, Obsidian Dome is a massive plug of volcanic glass, and some slopes in the Glass Mountains are pure obsidian. For Paiutes, these areas were invaluable for point material, and through trade, this obsidian spread over much of North America. Here, tourists of the 1920s play on an obsidian boulder near Obsidian Dome. (Courtesy of ECM.)

The east side of the Inyo Mountains is even steeper than the east side of the Sierra, and travel through a canyon can be a technically demanding adventure over cliffs, waterfalls, and through dense brush. At right, Will Crljenko rappels down Beveridge Canyon. (Courtesy of AS.)

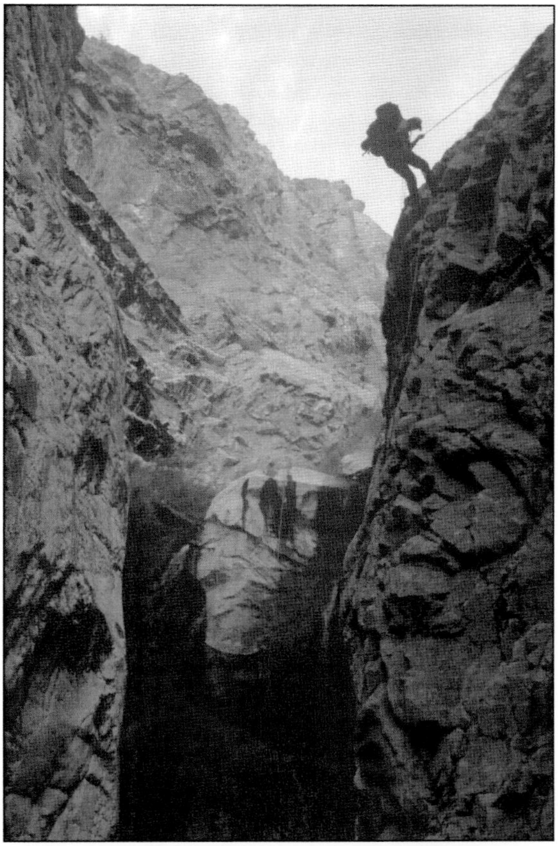

With no lakes and few reliable watercourses, the Inyo Mountains do not draw a lot of recreation. But for those, like George Lozito, who put some effort into hiking, back-road driving, and even skiing, there is great adventure amid quiet and magnificent scenery. (Courtesy of AS.)

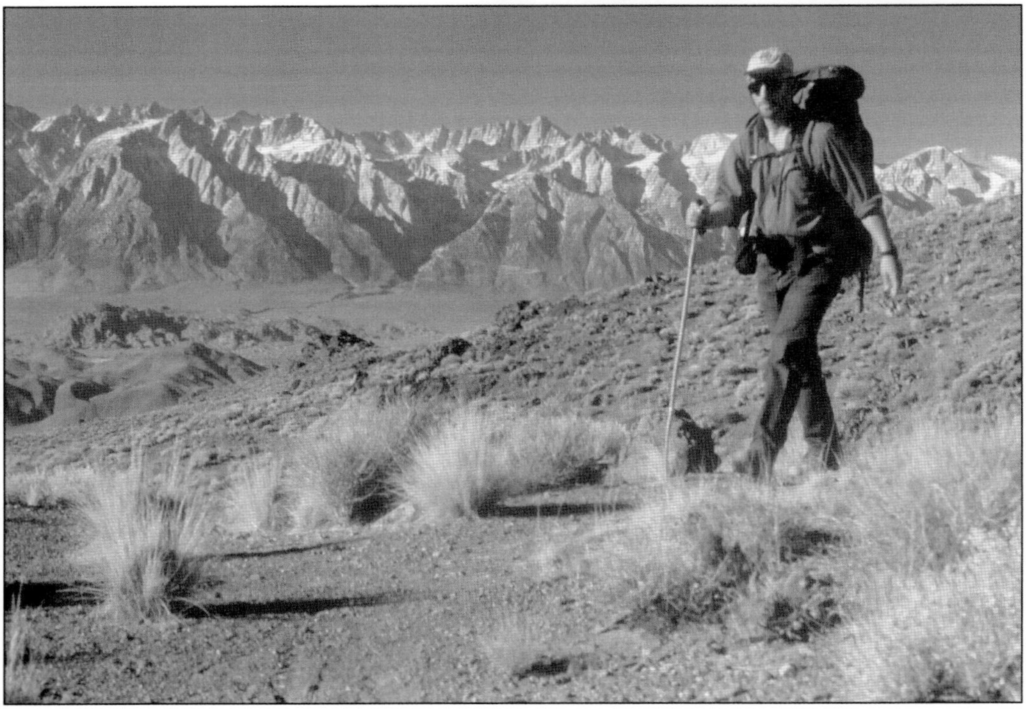

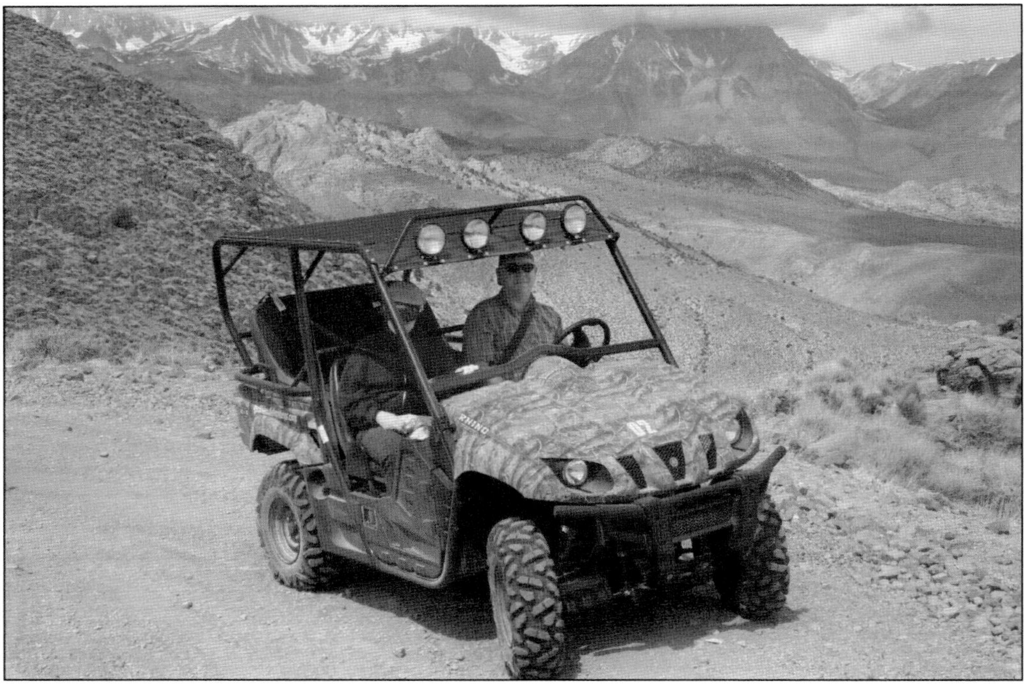

Automotive technology has developed specialized recreational vehicles for traveling the rockiest and sandiest backcountry roads. Here, riders on a modern all-terrain vehicle enjoy the rough trip up the Coyote Ridge, southwest of Bishop. (Courtesy of Randy Gillespie.)

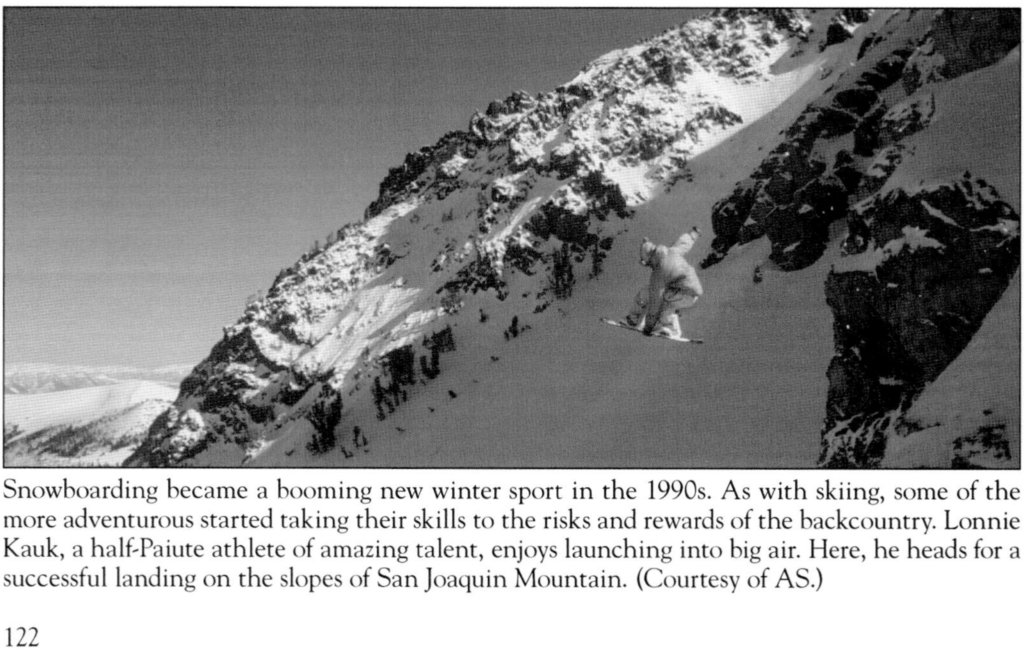

Snowboarding became a booming new winter sport in the 1990s. As with skiing, some of the more adventurous started taking their skills to the risks and rewards of the backcountry. Lonnie Kauk, a half-Paiute athlete of amazing talent, enjoys launching into big air. Here, he heads for a successful landing on the slopes of San Joaquin Mountain. (Courtesy of AS.)

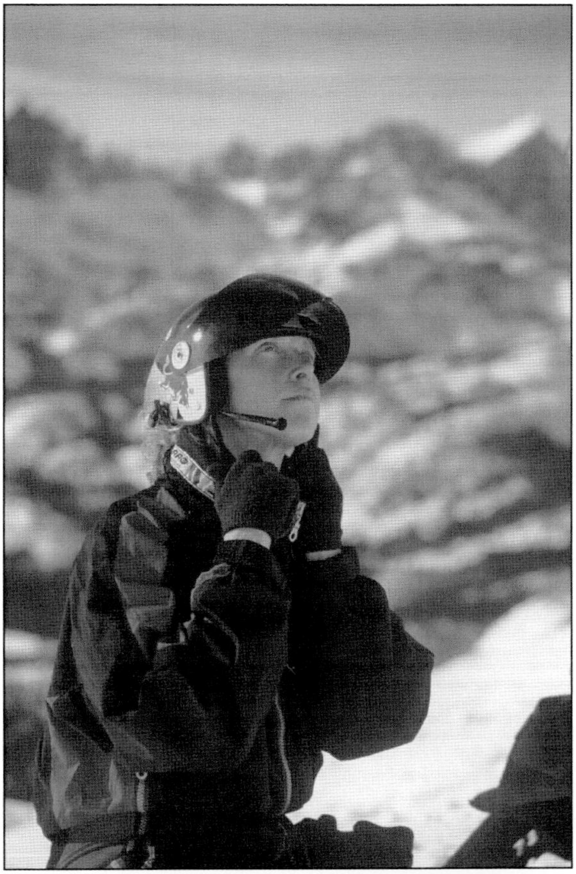

The great relief of the Whites and Sierra, combined with powerful daytime heating, creates tremendous atmospheric lift. Hang gliders and now paragliders have found that the Inyo is one of the best flying areas in the world. On good days, gliders can soar up to 20,000 feet, stay in the air for hours, and go over 100 miles. Paragliding champion Todd Bibler calls the Whites "the Himalayas of soaring." At right is Kari Castle, arguably the world's superstar of paragliding, who has made the Owens Valley her home for some 20 years. Below, a launch from the Whites gives gliders a great view of the Palisade peaks and glaciers. (Both images courtesy of AS.)

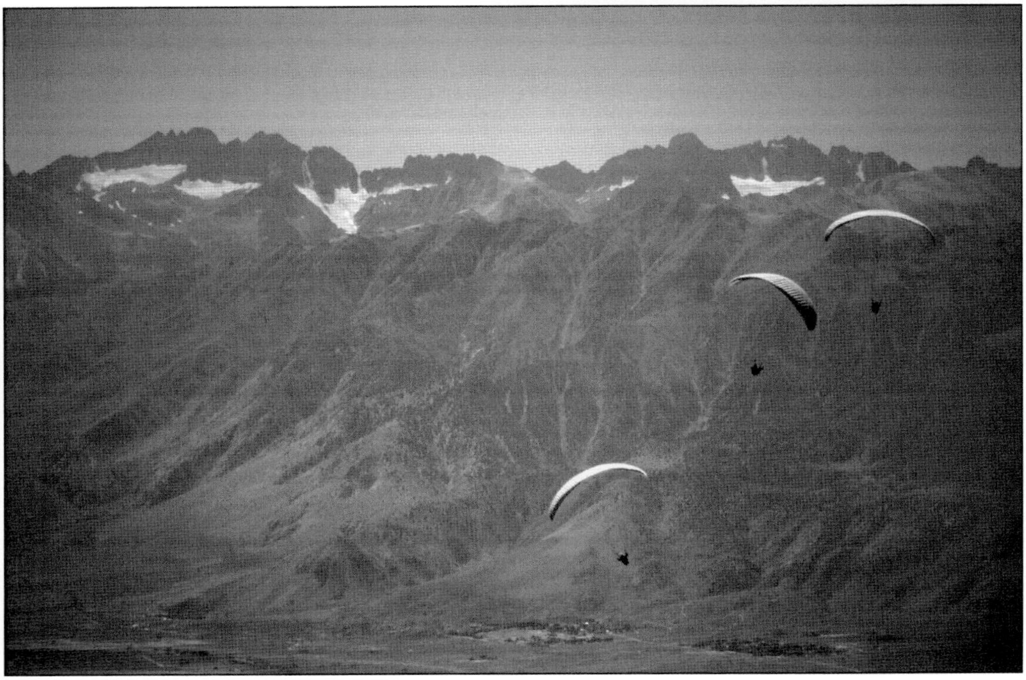

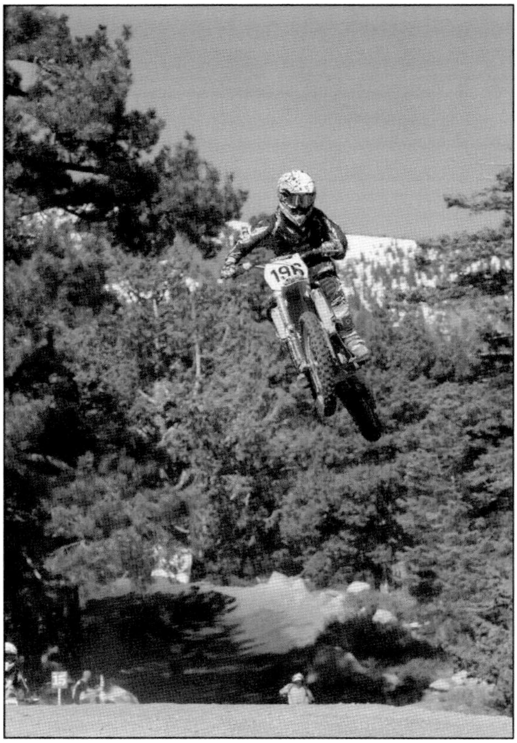

Motorcycles have long been a fun and efficient way to get around backcountry roads. In 1968, Dave McCoy got the go-ahead through district ranger Bill Murphy to build this motocross racecourse near Mammoth, beneath Sherwin Ridge. It remains the only motocross course on National Forest land, and is only open for event periods. (Courtesy of Susan Morning.)

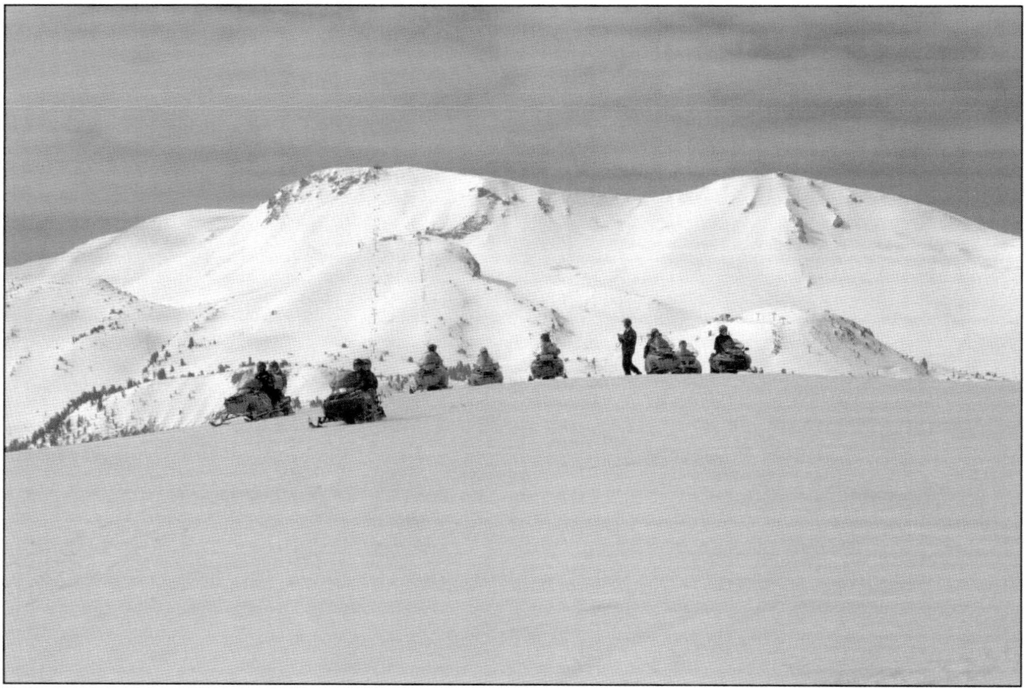

Snowmobiling sweeps riders over vast areas of winter whiteness at up to 60 miles per hour. Here, riders cruise with Mammoth Mountain as a backdrop. The ski area now operates a snowmobile rental and guide service, as does another company at Smokey Bear Flat. (Courtesy of Mammoth Mountain Ski Area.)

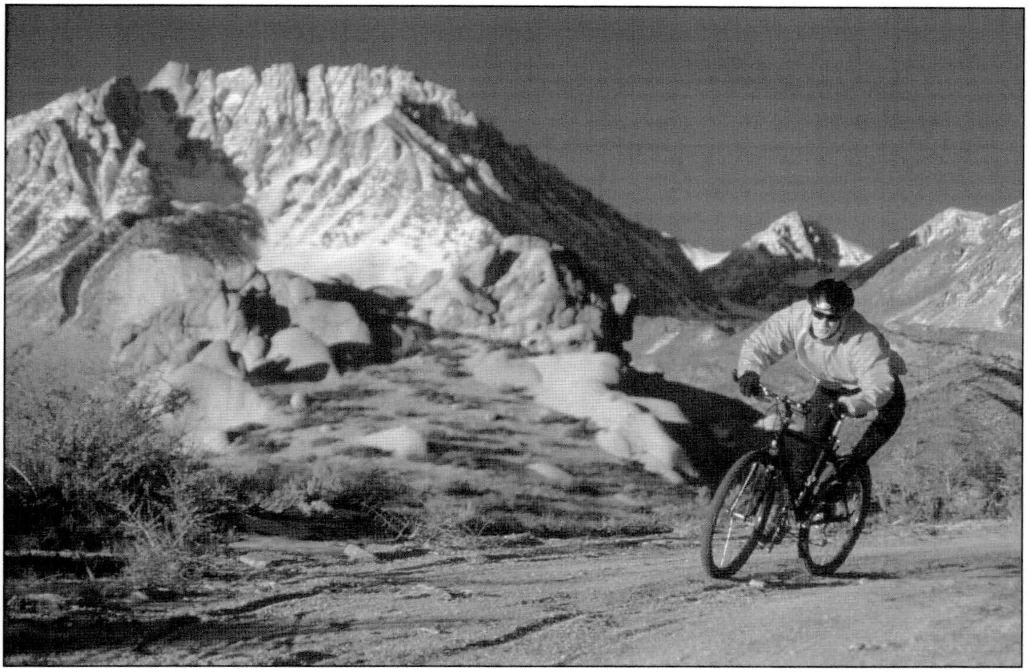

Mountain bikes became popular starting in the mid-1980s. Extensive dirt roads make the Inyo a popular destination for those who like to ride with concerns for self-sufficiency more than for traffic. For some, the downhills are the reason to ride, and Mammoth Mountain now has a system for summer riders to bring their bikes on the gondola and ride trails down. Here, Tove Jensen enjoys a downhill below Basin Mountain, above Bishop. (Courtesy of AS.)

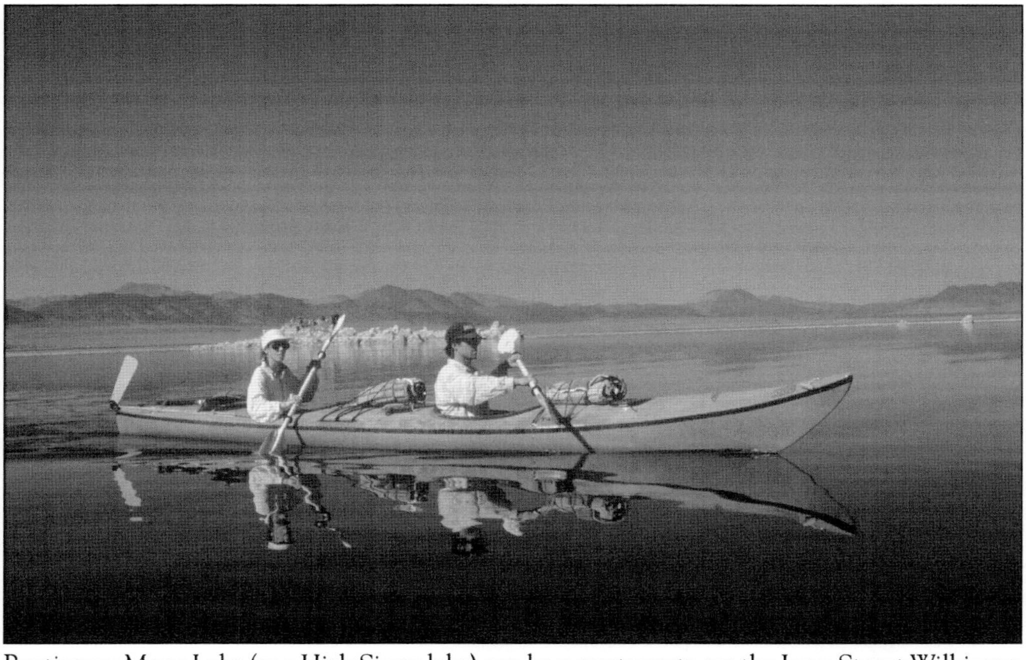

Boating on Mono Lake (or a High Sierra lake) can be a great way to see the Inyo. Stuart Wilkinson has been operating a kayak rental and guide business on the Inyo since about 1990. Here, he takes Carla Spencer on a tour of Mono Lake. (Courtesy of AS.)

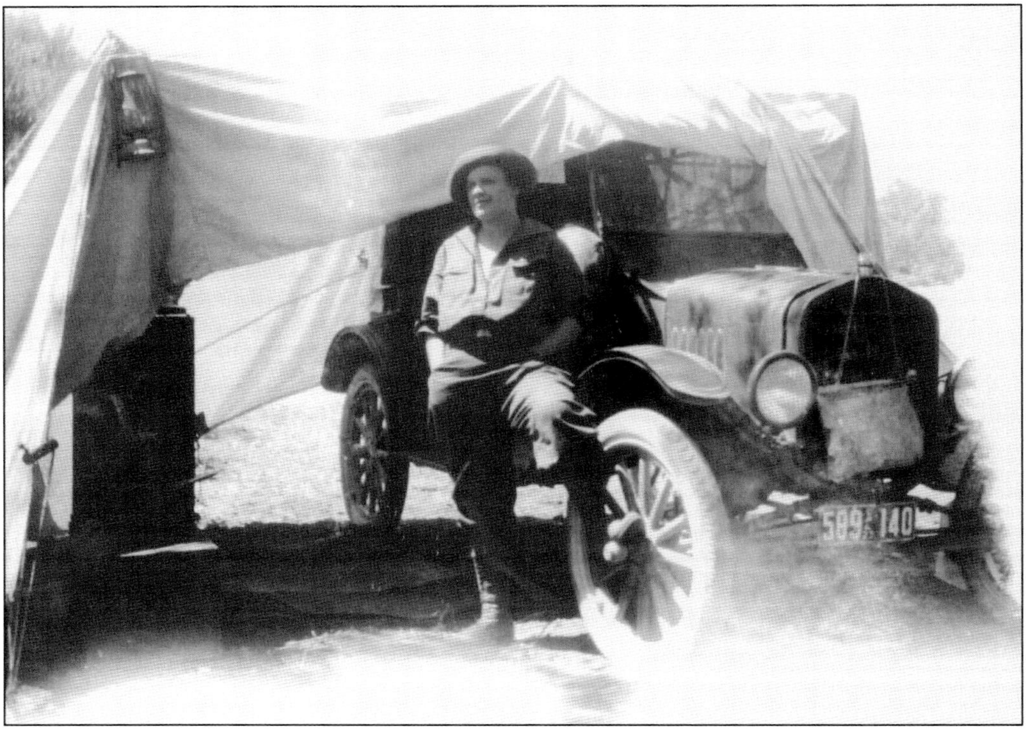

Of course, the automobile has given Americans great personal freedom. Long before there were any specialized vehicles for camping or driving on rough back roads, people used their cars to explore and camp. The knowledge of how to fix tires, crank-start, and do all sorts of other repairs was taken for granted. And, using one's vehicle as a half-support for a makeshift tent was common practice until manufacturers began selling outdoor gear more widely in the late 1920s. (Courtesy of INF.)

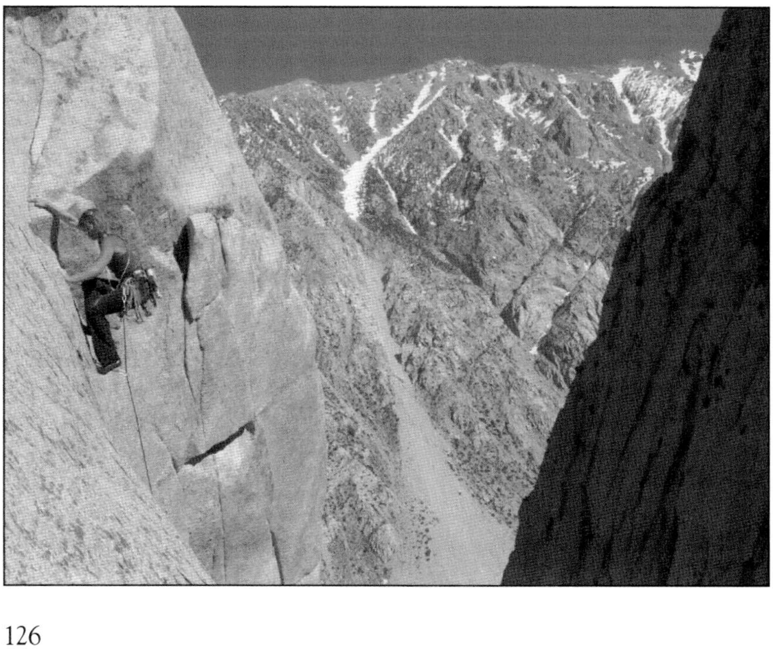

Exploring mountain terrain has become a fine art in many dimensions, and numerous crags along the Sierra canyons draw skilled and adventurous climbers to pick out dramatic new routes. Here, Melissa Buehler enjoys a demanding route in Box Canyon, a branch of Pine Creek. (Courtesy of AS.)

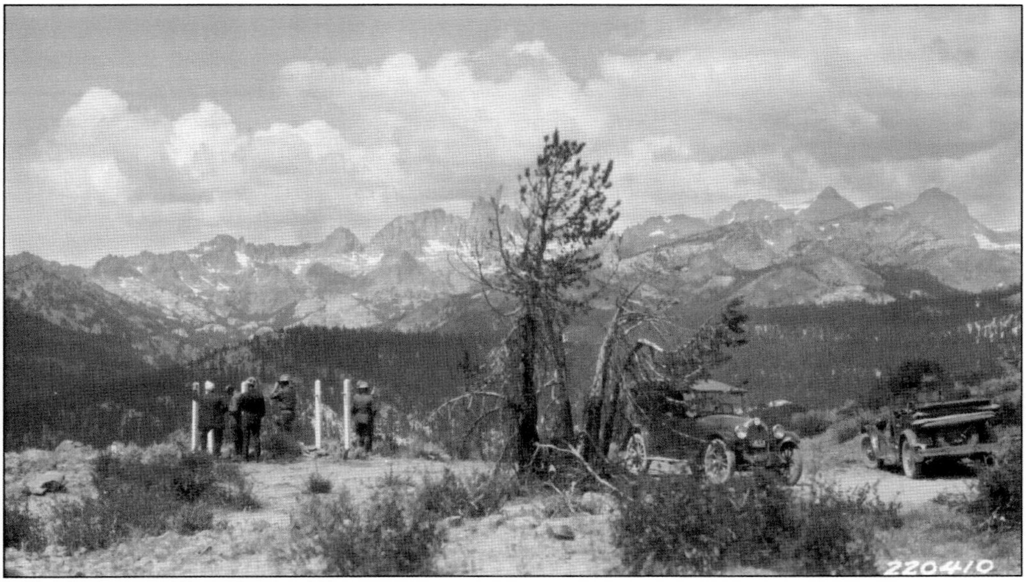

On the Inyo, just being in the scenery can be more than enough. Here, rangers in the mid-1920s take an early drive to Minaret Summit to install signs that interpret the panorama of the Ritter Range and upper San Joaquin River. (Courtesy of INF.)

About the Eastern Sierra Interpretive Association

The Eastern Sierra Interpretive Association was established in 1971 and is one of the region's oldest nonprofits. Today, ESIA's ongoing public land agency partnerships include the Inyo National Forest, Humboldt-Toiyabe National Forest, Tahoe National Forest, the Eastern Sierra InterAgency Visitor Center, and the Bureau of Land Management's Bishop Field Office. ESIA's primary mission is to assist its partners with educational, historical, scientific and interpretive activities. This is achieved through the operation of high quality bookstores located in Forest Service visitor centers and ranger stations selling educational books, maps and brochures. ESIA also sponsors special interpretive projects and programs. For more information visit www.esiaonline.com.

Eastern Sierra Interpretive Association